LONDON MATHEMATICAL SOCIETY LECTURE NOTE SERIES

Managing Editor: Professor Endre Süli, Mathematical Institute, University of Oxford,
Woodstock Road, Oxford OX2 6GG, United Kingdom

The titles below are available from booksellers, or from Cambridge University Press at
www.cambridge.org/mathematics

391 Fusion systems in algebra and topology, M. ASCHBACHER, R. KESSAR & B. OLIVER
392 Surveys in combinatorics 2011, R. CHAPMAN (ed)
393 Non-abelian fundamental groups and Iwasawa theory, J. COATES et al (eds)
394 Variational problems in differential geometry, R. BIELAWSKI, K. HOUSTON & M. SPEIGHT (eds)
395 How groups grow, A. MANN
396 Arithmetic differential operators over the p-adic integers, C.C. RALPH & S.R. SIMANCA
397 Hyperbolic geometry and applications in quantum chaos and cosmology, J. BOLTE & F. STEINER (eds)
398 Mathematical models in contact mechanics, M. SOFONEA & A. MATEI
399 Circuit double cover of graphs, C.-Q. ZHANG
400 Dense sphere packings: a blueprint for formal proofs, T. HALES
401 A double Hall algebra approach to affine quantum Schur–Weyl theory, B. DENG, J. DU & Q. FU
402 Mathematical aspects of fluid mechanics, J.C. ROBINSON, J.L. RODRIGO & W. SADOWSKI (eds)
403 Foundations of computational mathematics, Budapest 2011, F. CUCKER, T. KRICK, A. PINKUS & A. SZANTO (eds)
404 Operator methods for boundary value problems, S. HASSI, H.S.V. DE SNOO & F.H. SZAFRANIEC (eds)
405 Torsors, étale homotopy and applications to rational points, A.N. SKOROBOGATOV (ed)
406 Appalachian set theory, J. CUMMINGS & E. SCHIMMERLING (eds)
407 The maximal subgroups of the low-dimensional finite classical groups, J.N. BRAY, D.F. HOLT & C.M. RONEY-DOUGAL
408 Complexity science: the Warwick master's course, R. BALL, V. KOLOKOLTSOV & R.S. MACKAY (eds)
409 Surveys in combinatorics 2013, S.R. BLACKBURN, S. GERKE & M. WILDON (eds)
410 Representation theory and harmonic analysis of wreath products of finite groups, T. CECCHERINI-SILBERSTEIN, F. SCARABOTTI & F. TOLLI
411 Moduli spaces, L. BRAMBILA-PAZ, O. GARCÍA-PRADA, P. NEWSTEAD & R.P. THOMAS (eds)
412 Automorphisms and equivalence relations in topological dynamics, D.B. ELLIS & R. ELLIS
413 Optimal transportation, Y. OLLIVIER, H. PAJOT & C. VILLANI (eds)
414 Automorphic forms and Galois representations I, F. DIAMOND, P.L. KASSAEI & M. KIM (eds)
415 Automorphic forms and Galois representations II, F. DIAMOND, P.L. KASSAEI & M. KIM (eds)
416 Reversibility in dynamics and group theory, A.G. O'FARRELL & I. SHORT
417 Recent advances in algebraic geometry, C.D. HACON, M. MUSTAŢĂ & M. POPA (eds)
418 The Bloch–Kato conjecture for the Riemann zeta function, J. COATES, A. RAGHURAM, A. SAIKIA & R. SUJATHA (eds)
419 The Cauchy problem for non-Lipschitz semi-linear parabolic partial differential equations, J.C. MEYER & D.J. NEEDHAM
420 Arithmetic and geometry, L. DIEULEFAIT et al (eds)
421 O-minimality and Diophantine geometry, G.O. JONES & A.J. WILKIE (eds)
422 Groups St Andrews 2013, C.M. CAMPBELL et al (eds)
423 Inequalities for graph eigenvalues, Z. STANIĆ
424 Surveys in combinatorics 2015, A. CZUMAJ et al (eds)
425 Geometry, topology and dynamics in negative curvature, C.S. ARAVINDA, F.T. FARRELL & J.-F. LAFONT (eds)
426 Lectures on the theory of water waves, T. BRIDGES, M. GROVES & D. NICHOLLS (eds)
427 Recent advances in Hodge theory, M. KERR & G. PEARLSTEIN (eds)
428 Geometry in a Fréchet context, C.T.J. DODSON, G. GALANIS & E. VASSILIOU
429 Sheaves and functions modulo p, L. TAELMAN
430 Recent progress in the theory of the Euler and Navier–Stokes equations, J.C. ROBINSON, J.L. RODRIGO, W. SADOWSKI & A. VIDAL-LÓPEZ (eds)
431 Harmonic and subharmonic function theory on the real hyperbolic ball, M. STOLL
432 Topics in graph automorphisms and reconstruction (2nd Edition), J. LAURI & R. SCAPELLATO
433 Regular and irregular holonomic D-modules, M. KASHIWARA & P. SCHAPIRA
434 Analytic semigroups and semilinear initial boundary value problems (2nd Edition), K. TAIRA
435 Graded rings and graded Grothendieck groups, R. HAZRAT
436 Groups, graphs and random walks, T. CECCHERINI-SILBERSTEIN, M. SALVATORI & E. SAVA-HUSS (eds)
437 Dynamics and analytic number theory, D. BADZIAHIN, A. GORODNIK & N. PEYERIMHOFF (eds)
438 Random walks and heat kernels on graphs, M.T. BARLOW
439 Evolution equations, K. AMMARI & S. GERBI (eds)
440 Surveys in combinatorics 2017, A. CLAESSON et al (eds)

441	Polynomials and the mod 2 Steenrod algebra I,	G. WALKER & R.M.W. WOOD
442	Polynomials and the mod 2 Steenrod algebra II,	G. WALKER & R.M.W. WOOD
443	Asymptotic analysis in general relativity,	T. DAUDÉ, D. HÄFNER & J.-P. NICOLAS (eds)
444	Geometric and cohomological group theory,	P.H. KROPHOLLER, I.J. LEARY, C. MARTÍNEZ-PÉREZ & B.E.A. NUCINKIS (eds)
445	Introduction to hidden semi-Markov models,	J. VAN DER HOEK & R.J. ELLIOTT
446	Advances in two-dimensional homotopy and combinatorial group theory,	W. METZLER & S. ROSEBROCK (eds)
447	New directions in locally compact groups,	P.-E. CAPRACE & N. MONOD (eds)
448	Synthetic differential topology,	M.C. BUNGE, F. GAGO & A.M. SAN LUIS
449	Permutation groups and cartesian decompositions,	C.E. PRAEGER & C. SCHNEIDER
450	Partial differential equations arising from physics and geometry,	M. BEN AYED et al (eds)
451	Topological methods in group theory,	N. BROADDUS, M. DAVIS, J.-F. LAFONT & I. ORTIZ (eds)
452	Partial differential equations in fluid mechanics,	C.L. FEFFERMAN, J.C. ROBINSON & J.L. RODRIGO (eds)
453	Stochastic stability of differential equations in abstract spaces,	K. LIU
454	Beyond hyperbolicity,	M. HAGEN, R. WEBB & H. WILTON (eds)
455	Groups St Andrews 2017 in Birmingham,	C.M. CAMPBELL et al (eds)
456	Surveys in combinatorics 2019,	A. LO, R. MYCROFT, G. PERARNAU & A. TREGLOWN (eds)
457	Shimura varieties,	T. HAINES & M. HARRIS (eds)
458	Integrable systems and algebraic geometry I,	R. DONAGI & T. SHASKA (eds)
459	Integrable systems and algebraic geometry II,	R. DONAGI & T. SHASKA (eds)
460	Wigner-type theorems for Hilbert Grassmannians,	M. PANKOV
461	Analysis and geometry on graphs and manifolds,	M. KELLER, D. LENZ & R.K. WOJCIECHOWSKI
462	Zeta and L-functions of varieties and motives,	B. KAHN
463	Differential geometry in the large,	O. DEARRICOTT et al (eds)
464	Lectures on orthogonal polynomials and special functions,	H.S. COHL & M.E.H. ISMAIL (eds)
465	Constrained Willmore surfaces,	Á.C. QUINTINO
466	Invariance of modules under automorphisms of their envelopes and covers,	A.K. SRIVASTAVA, A. TUGANBAEV & P.A. GUIL ASENSIO
467	The genesis of the Langlands program,	J. MUELLER & F. SHAHIDI
468	(Co)end calculus,	F. LOREGIAN
469	Computational cryptography,	J.W. BOS & M. STAM (eds)
470	Surveys in combinatorics 2021,	K.K. DABROWSKI et al (eds)
471	Matrix analysis and entrywise positivity preservers,	A. KHARE
472	Facets of algebraic geometry I,	P. ALUFFI et al (eds)
473	Facets of algebraic geometry II,	P. ALUFFI et al (eds)
474	Equivariant topology and derived algebra,	S. BALCHIN, D. BARNES, M. KEDZIOREK & M. SZYMIK (eds)
475	Effective results and methods for Diophantine equations over finitely generated domains,	J.-H. EVERTSE & K. GYŐRY
476	An indefinite excursion in operator theory,	A. GHEONDEA
477	Elliptic regularity theory by approximation methods,	E.A. PIMENTEL
478	Recent developments in algebraic geometry,	H. ABBAN, G. BROWN, A. KASPRZYK & S. MORI (eds)
479	Bounded cohomology and simplicial volume,	C. CAMPAGNOLO, F. FOURNIER-FACIO, N. HEUER & M. MORASCHINI (eds)
480	Stacks Project Expository Collection (SPEC),	P. BELMANS, W. HO & A.J. DE JONG (eds)
481	Surveys in combinatorics 2022,	A. NIXON & S. PRENDIVILLE (eds)
482	The logical approach to automatic sequences,	J. SHALLIT
483	Rectifiability: a survey,	P. MATTILA
484	Discrete quantum walks on graphs and digraphs,	C. GODSIL & H. ZHAN
485	The Calabi problem for Fano threefolds,	C. ARAUJO et al
486	Modern trends in algebra and representation theory,	D. JORDAN, N. MAZZA & S. SCHROLL (eds)
487	Algebraic combinatorics and the Monster group,	A.A. IVANOV (ed)
488	Maurer–Cartan methods in deformation theory,	V. DOTSENKO, S. SHADRIN & B. VALLETTE
489	Higher dimensional algebraic geometry,	C. HACON & C. XU (eds)
490	C^∞-algebraic geometry with corners,	K. FRANCIS-STAITE & D. JOYCE
491	Groups and graphs, designs and dynamics,	R.A. BAILEY, P.J. CAMERON & Y. WU (eds)
492	Homotopy theory of enriched Mackey functors,	N. JOHNSON & D. YAU
493	Surveys in combinatorics 2024,	F. FISCHER & R. JOHNSON (eds)
494	K-theory and representation theory,	R. PLYMEN & M.H. ŞENGÜN (eds)
495	Polygraphs: from rewriting to higher categories,	D. ARA et al
496	Groups St Andrews 2022 in Newcastle,	C.M. CAMPBELL et al (eds)
497	Proof complexity generators,	J. KRAJÍČEK
498	Polynomial functors,	N. NIU & D.I. SPIVAK
499	Moduli, motives and bundles,	P.L. DEL ÁNGEL R., F. NEUMANN & A.H.W. SCHMITT (eds)
500	The Toda lattice and universality for the computation of the eigenvalues of a random matrix,	P. DEIFT, G. DUBACH, C. TOMEI & T. TROGDON

London Mathematical Society Lecture Note Series: 500

The Toda Lattice and Universality for the Computation of the Eigenvalues of a Random Matrix

PERCY DEIFT
New York University

GUILLAUME DUBACH
École Polytechnique, Paris

CARLOS TOMEI
Pontifical Catholic University of Rio de Janeiro (PUC-Rio)

THOMAS TROGDON
University of Washington

Shaftesbury Road, Cambridge CB2 8EA, United Kingdom

One Liberty Plaza, 20th Floor, New York, NY 10006, USA

477 Williamstown Road, Port Melbourne, VIC 3207, Australia

314–321, 3rd Floor, Plot 3, Splendor Forum, Jasola District Centre,
New Delhi – 110025, India

103 Penang Road, #05–06/07, Visioncrest Commercial, Singapore 238467

Cambridge University Press is part of Cambridge University Press & Assessment,
a department of the University of Cambridge.

We share the University's mission to contribute to society through the pursuit of
education, learning and research at the highest international levels of excellence.

www.cambridge.org
Information on this title: www.cambridge.org/9781009664356
DOI: 10.1017/9781009664332

© Percy Deift, Guillaume Dubach, Carlos Tomei and Thomas Trogdon 2026

This publication is in copyright. Subject to statutory exception and to the provisions
of relevant collective licensing agreements, no reproduction of any part may take
place without the written permission of Cambridge University Press & Assessment.

When citing this work, please include a reference to the DOI 10.1017/9781009664332

First published 2026

A catalogue record for this publication is available from the British Library

*A Cataloging-in-Publication data record for this book is available from the
Library of Congress*

ISBN 978-1-009-66435-6 Paperback

Cambridge University Press & Assessment has no responsibility for the persistence
or accuracy of URLs for external or third-party internet websites referred to in this
publication and does not guarantee that any content on such websites is, or will
remain, accurate or appropriate.

For EU product safety concerns, contact us at Calle de José Abascal, 56, 1o, 28003
Madrid, Spain, or email eugpsr@cambridge.org

This monograph is dedicated to the memory of H. Flaschka (1945–2021), S.V. Manakov (1948–2012) and J.K. Moser (1928–1999), for their profound contributions to the study of integrable systems and, in particular, of the Toda lattice.

Contents

		Preface	*page* ix
1		**Introduction**	1
	1.1	A Case Study: Eigenvalue Computation	1
	1.2	Description of the Algorithms	5
	1.3	Goal of This Work	11
	1.4	Outline for the Monograph	12
	1.5	Related Scenarios and Some Notational Issues	13
2		**Hamiltonian Mechanics and Integrable Systems**	16
	2.1	Hamiltonians on Symplectic Manifolds	16
	2.2	How Symplectic Manifolds Arise	20
	2.3	Integration of Vector Fields	26
	2.4	Two Classical Examples	34
3		**The Toda Lattice**	37
	3.1	The Tridiagonal Case	37
	3.2	Long-Time Behavior	46
	3.3	Liouville Integrability of the Toda Lattice	62
	3.4	The Toda Lattice: Full Matrix Case	65
	3.5	Long-Time Behavior in the Full Hermitian Case	72
	3.6	The Generalized Toda Flow	75
	3.7	Convergence to the Top Eigenvalue	80
	3.8	Action-Angle Variables for the Toda Flow on Jacobi Matrices	86
	3.9	Action-Angle Variables for Toda Flow on Full Symmetric Matrices	96
	3.10	The Stroboscope Theorem	100
4		**Toda without Hamiltonian Structure**	104
	4.1	Another Set of Linearizing Variables	104

	4.2	Profiles and Isospectral Manifolds	110
	4.3	The Tridiagonal Isospectral Manifold	118
5	**Random Matrix Ensembles**		122
	5.1	Invariant and Wigner Ensembles	122
	5.2	Estimates from Random Matrix Theory	123
	5.3	Technical Lemmas	127
6	**Universality for the Toda Algorithm**		129
	6.1	A Numerical Demonstration	133
	6.2	Estimates for the Toda Algorithm	137
	6.3	Adding Probability	145
	References		152
	Notations and Abbreviations		157
	Index		160

Preface

This monograph is based on a course given by P. Deift at the Courant Institute in the Spring term of 2019. Its main goal is to analyze the behavior of the Toda algorithm for computing the eigenvalues of a random real symmetric or Hermitian matrix. Of particular interest is the statistics of the stopping times of the algorithm for computing the eigenvalues to a given accuracy ϵ. It turns out that these statistics, suitably centered and scaled, are universal, in the sense that they are independent of the choice of a very broad class of random matrix ensembles, which includes Wigner Ensembles and Invariant Ensembles. Universality of stopping times for numerical algorithms with random data has now been observed for a large variety of numerical computations. It is the contention of the authors that universality with respect to random data is a new and general phenomenon in computation, of which the Toda algorithm is just a particularly striking example.

For the convenience of the reader who may not be familiar with the Toda flow, which underlies the Toda algorithm, we include a detailed analysis of the flow, especially a description of the asymptotic behavior of its solutions. Most of these details are well known, but some are perhaps less familiar, even to experts in the theory of integrable systems. For example, the monograph contains a novel presentation of the Toda lattice as a dynamical system stripped of Hamiltonian structure: the explicit solvability of the lattice now results, not from Liouville's Theorem, but rather from a simple interplay of Gaussian elimination and the Gram–Schmidt orthogonalization process for independent vectors in a Hilbert space.

The authors would like to thank Yuri Bakhtin and Paul Bourgade for many useful conversations, and Folkmar Bornemann for the data on F_2^{gap}. We also thank Jeanne Boursier, Emmanuel Memmi, Victor Dubach, and Margaret Bilu for their help in editing and proofreading the manuscript. Thanks also to the reviewers of the manuscript for their insightful and useful comments. Finally, the authors are grateful to David Tranah of Cambridge University Press for his patience as we delayed the manuscript again and again, and for his many very useful suggestions to improve the text. Thank you, David.

This work was supported in part by grants NSF-DMS-1303018, NSF-DMS-1945652, NSF-DMS-2306438 (TT), NSF-DMS-1300965 (PD), StoneLab, FAPERJ E-26/200-980/2022 and CNPq 304742/2021-0 (CT). P. Deift acknowledges the support of a Silver Grant at New York University. G. Dubach gratefully acknowledges support from the grants NSF DMS-1812114 of P. Bourgade and NSF CAREER DMS-1653602 of L.-P. Arguin.

1
Introduction

1.1 A Case Study: Eigenvalue Computation

In the early 2010s, Christian Pfrang, Govind Menon and Percy Deift initiated a statistical study of the performance of various standard algorithms for computing the eigenvalues of random real symmetric matrices (Pfrang et al., 2014). The eigenvalues of an $N \times N$ matrix M comprise the *spectrum* of M which we denote by $\text{spec}(M)$, or, sometimes, $\sigma(M)$.

Let Σ_N denote the set of $N \times N$ *real symmetric matrices*. Associated with each discrete algorithm $\mathcal{A}$ of interest, there is a map

$$\varphi_{\mathcal{A}} : \Sigma_N \to \Sigma_N$$

with the properties

- *Isospectrality* : $\text{spec}(\varphi_{\mathcal{A}}(H)) = \text{spec}(H)$, for any $H \in \Sigma_N$.
- *Convergence* : the iterates

$$X_{k+1} = \varphi_{\mathcal{A}}(X_k), \quad k \geq 0,$$

with $X_0 = H$ given, converge generically to a diagonal matrix,

$$X_k \xrightarrow[k \to \infty]{} X_\infty.$$

In the continuous case, there is a flow $t \mapsto X(t) \in \Sigma_N$ with similar properties

- *Isospectrality* : $\text{spec}(X(t))$ is constant.
- *Convergence* : the flow $X(t)$ with initial condition $X(0) = H$ converges to a diagonal matrix,

$$X(t) \xrightarrow[t \to \infty]{} X_\infty.$$

In both cases, necessarily, the diagonal entries of X_∞ are the eigenvalues λ_i of the given matrix H.

Given $\varepsilon > 0$, it follows, in the discrete case, that for some m, the off-diagonal entries of X_m are $O(\varepsilon)$, and hence the diagonal entries of X_m give the eigenvalues of H up to $O(\varepsilon)$. The situation is similar for continuous algorithms $t \mapsto X(t)$. However, rather than running the algorithm until all the off-diagonal entries are $O(\varepsilon)$, it is customary to run the algorithm with deflations, as follows.

For an $N \times N$ matrix Y in block form

$$Y = \begin{pmatrix} Y_{11} & Y_{12} \\ Y_{21} & Y_{22} \end{pmatrix}$$

with Y_{11} of size $k \times k$ and Y_{22} of size $(N-k) \times (N-k)$ for some $k \in \{1, \ldots, N-1\}$, the process of projecting $Y \mapsto \text{diag}(Y_{11}, Y_{22})$ is called a *deflation*. For a given algorithm $\mathcal{A}$, $\varepsilon > 0$ and $H \in \Sigma_N$, we define the k-deflation time

$$T^{(k)}(H) = T_{\varepsilon,\mathcal{A}}^{(k)}, \qquad 1 \le k \le N-1,$$

to be the smallest value of m such that X_m, the mth iterate of the algorithm $\mathcal{A}$ applied to $X_0 = H$, has block form

$$X_m = \varphi_\mathcal{A}^m(X_0) = \begin{pmatrix} X_{11}^{(k)} & X_{12}^{(k)} \\ X_{21}^{(k)} & X_{22}^{(k)} \end{pmatrix}$$

with $X_{11}^{(k)}$ of size $k \times k$, $X_{22}^{(k)}$ of size $(N-k) \times (N-k)$ and

$$\|X_{12}^{(k)}\| = \|X_{21}^{(k)}\| \le \varepsilon.$$

The deflation time $T(H)$ is then defined as

$$T(H) = T_{\varepsilon,\mathcal{A}}(H) = \min_{1 \le k \le N-1} T_{\varepsilon,\mathcal{A}}^{(k)}(H).$$

If $k \in \{1, \ldots, N-1\}$ is such that $T(H) = T_{\varepsilon,\mathcal{A}}^{(k)}(H)$, the eigenvalues of $H = X_0$ are given by the eigenvalues of a block-diagonal matrix $\text{diag}(X_{11}^{(k)}, X_{22}^{(k)})$ up to an error $O(\varepsilon)$. After running the algorithm to time $T_{\varepsilon,\mathcal{A}}(H)$, the algorithm restarts by applying the basic algorithm $\mathcal{A}$ separately to the smaller matrices $X_{11}^{(k)}$ and $X_{22}^{(k)}$ until the next deflation time, and so on. There are again similar considerations for continuous algorithms.

As the algorithm proceeds, the number of matrices after each deflation doubles. This is counterbalanced by the fact that the matrices are smaller and smaller in size, and the calculations are clearly parallelizable. Allowing for parallel computation, the number of deflations to compute all the

1.1 A Case Study: Eigenvalue Computation

eigenvalues of a given matrix H to a given accuracy ε, will vary from $O(\ln N)$ to $O(N)$.

A prototype example of a discrete algorithm is the so-called QR algorithm (see, e.g., Golub and Loan, 2013) and an example of a continuous algorithm is the so-called Toda algorithm (see, e.g., Deift et al., 1983). We will describe these algorithms in detail below.

In Pfrang et al. (2014), the authors considered the deflation time $T_{\varepsilon,\mathcal{A}}$ for $N \times N$ matrices chosen from a given ensemble $\mathcal{E}$. Henceforth we suppress the dependence on $\varepsilon, N, \mathcal{A}$ and $\mathcal{E}$ and simply write T with these variables understood. For a given algorithm $\mathcal{A}$ and ensemble $\mathcal{E}$, the authors computed $T(H)$ for 5,000 to 15,000 samples of matrices H chosen from $\mathcal{E}$, and recorded the **normalized deflation time**

$$\widetilde{T}(H) := \frac{T(H) - \langle T \rangle}{\sigma},$$

where $\langle T \rangle$ and $\sigma^2 = \langle (T - \langle T \rangle)^2 \rangle$ are the sample average and sample variance of $T(H)$, respectively. Surprisingly, the authors found that for the given algorithm $\mathcal{A}$, and ε, N in a suitable scaling range as $N \to \infty$, the histogram of $\widetilde{T}$ was **universal**. In other words, the fluctuations in the deflation time $\widetilde{T}$, scaled as above, were **independent** of the ensemble, for a wide range of ensembles $\mathcal{E}$. Figure 1.1 displays, in a slightly different form, some of the numerical results from Pfrang et al. (2014). Panel (a) displays data for the (discrete) QR algorithm, whereas panel (b) displays data for the (continuous) Toda algorithm.

Subsequently, Govind Menon, Sheehan Olver, Thomas Trogdon, and Percy Deift (2014) raised the question of whether the universality results of Pfrang et al. (2014) were limited to eigenvalue algorithms for real symmetric matrices, or whether they were present more generally in numerical computation. And indeed, they found similar universality results for a wide variety of numerical algorithms, including

(a) other algorithms such as the QR algorithm with shifts,[1] the Jacobi eigenvalue algorithm, and also algorithms applied to complex Hermitian ensembles.
(b) The conjugate gradient and GMRES algorithms to solve linear systems $Hx = b$, with H and b random.
(c) An iterative algorithm to solve the Dirichlet problem $\Delta u = 0$ on a random star-shaped region $\Omega \subset \mathbb{R}^2$ with random boundary data f on $\partial \Omega$, and

---
[1] The QR algorithm with shifts is the accelerated version that is used in practice.

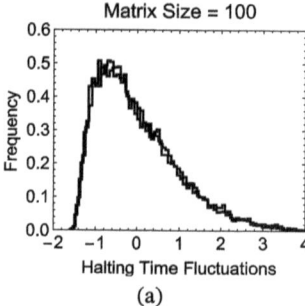
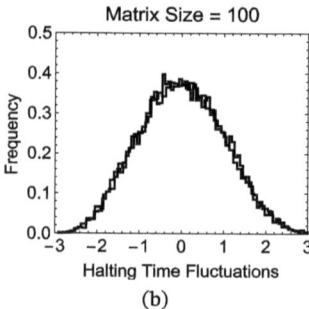

Figure 1.1 Universality for $\tilde{T}$ when (a) $\mathcal{A}$ is the QR eigenvalue algorithm and when (b) $\mathcal{A}$ is the Toda algorithm. Panel (a) displays the overlay of two histograms for $\tilde{T}$ in the case of QR, one for each of the two ensembles $\mathcal{E}$ = BE, consisting of i.i.d. mean-zero Bernoulli random variables, and $\mathcal{E}$ = GOE, consisting of i.i.d. mean-zero normal random variables. Here $\varepsilon = 10^{-10}$ and $N = 100$. Panel (b) displays the overlay of two histograms for $\tilde{T}$ in the case of the Toda algorithm, and again $\mathcal{E}$ = BE or GOE. And here $\varepsilon = 10^{-8}$ and $N = 100$.

(d) A genetic algorithm[2] to compute the equilibrium measure for orthogonal polynomials on the line.

In Deift et al. (2014), the authors also discussed similar universality results obtained by Bakhtin and Correll (2012) in a series of experiments with live participants, recording the

(e) decision-making times for a specific task.

Whereas (a) and (b) concern problems of finite dimensions, (c) shows that universality is also present in infinite-dimensional settings. And whereas (a), (b) and (c) concern, in effect, deterministic dynamical systems acting on random initial data, problem (d) shows that universality is also present in genuinely stochastic algorithms.

The demonstration of universality in problems (a) to (d) raises the following issue: Given the common view of neuroscientists that the brain is adequately modelled as a "big computer" with hardware and software, one should be able to find evidence of universality in some neural computations. It is this issue that led Deift et al. (2014) to the work of Bakhtin and Correll.

---

[2] A genetic algorithm is an iterative procedure, inspired by Darwin's theory of natural selection, used to find solutions to optimization and search problems in high dimension. It is typically based on elementary operations metaphorically called "mutation," "crossover," or "selection."

All of the above results are numerical or experimental. In order to establish universality as a *bona fide* phenomenon in numerical analysis, and not just an artifact suggested, however strongly, by certain computations as above, Deift and Trogdon sought out and proved universality for a particular algorithm of interest. The algorithm they analyzed in Deift and Trogdon (2018) was the Toda eigenvalue algorithm to compute the top eigenvalue of a random real symmetric (alternatively Hermitian) matrix. The goal of this monograph is to describe this work in detail. In the subsequent work (Deift and Trogdon, 2017), the authors also proved universality for other eigenvalue algorithms, including QR on sample covariance matrices.

1.2 Description of the Algorithms

We now describe the Toda and QR algorithms. Here we focus on the case where the matrices are real symmetric, i.e., $H \in \Sigma_N$. The complex Hermitian case will be considered in Sections 3.5 and 3.6.

1.2.1 The Toda Flow

The Toda equations have the *Lax pair* form

$$\partial_t X = [X, B(X)] = XB(X) - B(X)X, \qquad X(0) = H, \qquad (1.1)$$

where

$$B(X) = X_- - X_-^T, \qquad (1.2)$$

X_- being the strictly lower-triangular part of X.

Theorem 1.1 *The Toda flow has a globally defined solution $X(t) \in \Sigma_N$. The solution admits two representations by conjugation,*

$$X(t) = Q(t)^T H Q(t) = R(t) H R(t)^{-1},$$

where $Q(t)$ are orthogonal matrices and $R(t)$ are upper triangular matrices with positive diagonal entries, $Q(0) = R(0) = I$. The eigenvalues are conserved quantities: The spectra of $X(t)$ and $X(0) = H$ are equal.

Proof The equations clearly satisfy the standard Cauchy-Lipschitz conditions, and hence have a unique local solution $X(t)$ for some $t^* \in (0, +\infty]$ and $t \in [0, t^*)$. Also, as $B(X) = -B(X)^T$, it follows that if X is symmetric,

$$[X, B(X)]^T = [X, B(X)].$$

Thus $[X, B(X)]$ defines a vector field on Σ_N, and hence as $X(0) = X^T(0) = H$, we must have

$$X(t) = X^T(t), \qquad 0 \le t < t^*, \tag{1.3}$$

i.e., the Toda equations preserve Σ_N. On this time interval,

$$\partial_t(X^2) = (\partial_t X)X + X(\partial_t X) = (XB - BX)X + X(XB - BX) = [X^2, B(X)],$$

i.e.,

$$\partial_t X^2 = [X^2, B(X)], \qquad 0 \le t < t^*. \tag{1.4}$$

As the trace of a commutator is always zero, we have $\partial_t(\operatorname{tr} X^2) = 0$, so that

$$\operatorname{tr} X^2(t) = \operatorname{tr} H^2. \tag{1.5}$$

But, as $X = X^T$, this implies

$$\sum_{i,j=1}^{N} X_{ij}^2(t) = \sum_{i,j=1}^{N} H_{ij}^2. \tag{1.6}$$

In particular,

$$|X_{ij}| \le \sqrt{\operatorname{tr} H^2} < \infty, \tag{1.7}$$

which gives an a priori bound on the entries of $X(t)$. It follows, by standard ODE techniques, that in fact $t^* = +\infty$, i.e., the equation has a unique global solution $X(t)$.

Let $Q(t), t \ge 0$, be the solution of the equation

$$\partial_t Q(t) = Q(t) B(X(t)), \qquad Q(0) = I, \tag{1.8}$$

where $X(t)$ solves (1.1). As (1.8) is linear, the solution is unique and global. Then, we have

$$\partial_t(QQ^T) = \partial_t QQ^T + Q\partial_t Q^T = QBQ^T + Q(-B)Q^T = 0$$

and so

$$Q(t)Q(t)^T = I,$$

i.e., $Q(t)$ is orthogonal for all $t \ge 0$. Now set

$$\widetilde{X}(t) := Q(t)^T H Q(t). \tag{1.9}$$

Then

$$\partial_t \widetilde{X} = \partial_t Q^T H Q + Q^T H \partial_t Q = -BQ^T H Q + Q^T H Q B = [\widetilde{X}, B(X)] \tag{1.10}$$

so that $\widetilde{X}$ solves the same linear equation as $X(t)$ (with $B(X) = B(X(t))$ given), and as $\widetilde{X}(0) = X(0)$, we must have $\widetilde{X}(t) = X(t)$. Thus

$$X(t) = Q(t)^T H Q(t). \tag{1.11}$$

So $X(t)$ is orthogonally equivalent to $X(0) = H$. In particular the flow $t \mapsto X(t)$ is isospectral: The eigenvalues of $X(0)$ and $X(t)$ are the same.

Finally, rewrite equation (1.1),

$$\partial_t X = [X, B(X)] = [X, B(X) - X] = [X, -U(X)], \quad X(0) = H, \tag{1.12}$$

where $U(X) = X - B(X)$ is an upper triangular matrix. Imitating the argument beginning at (1.8), consider the solution of equation

$$\partial_t R(t) = U(X(t)) R(t), \quad R(0) = I, \tag{1.13}$$

to obtain the alternative formula

$$X(t) = R(t) H R(t)^{-1}. \tag{1.14}$$

The evolution of the diagonal entries of $R(t)$ is given by

$$\partial_t R_{ii}(t) = (U(X(t)))_{ii} R(t)_{ii}.$$

As $R_{ii}(0) = 1 \neq 0$, we must have $R_{ii}(t) > 0$ for $t \in \mathbb{R}$. □

In Chapter 3, we use an argument from Moser (1975b) to show that indeed

$$X(t) \xrightarrow[t \to \infty]{} X_\infty, \tag{1.15}$$

where

$$X_\infty = \operatorname{diag}(\lambda_1, \ldots, \lambda_N). \tag{1.16}$$

As noted earlier, the λ_i are the eigenvalues of $X(0) = H$. We will also show that, generically, the flow (1.1) is sorting, i.e., $\lambda_1 \geq \cdots \geq \lambda_N$.

The Toda flow (1.1) gives rise to an eigenvalue algorithm in which, in order to compute the eigenvalues of a given matrix $H \in \Sigma_N$, one solves (1.1) with initial condition $X(0) = H$. Then,

$$X(t) \xrightarrow[t \to +\infty]{} \operatorname{diag}(\lambda_1, \ldots, \lambda_N),$$

where the λ_i are the eigenvalues of H. This is the *Toda algorithm* suggested by Kostant (1979) and independently by Deift et al. (1983), and analyzed in detail therein. Further discussion of the Toda algorithm and related algorithms can be found in Watkins (1984) and Chu (1984). We show in Chapter 3 that (1.1) is an integrable Hamiltonian flow in the sense of Liouville, which can be integrated explicitly. In Chapter 4, we do more: We show that the algorithm is *super-integrable* for a large class of subsets of Σ_N, the so called matrices with given *profile* (see Chapter 4), which includes the vector space of tridiagonal matrices.

The history of the Toda system, or Toda lattice, is as follows.[3] The system was introduced by H. Toda in 1967 and, after rescaling and a change of variables, it describes the motion of an infinite system of particles on the line under the Hamiltonian

$$H_{\text{Toda}}(x, y) = \frac{1}{2} \sum_{i=-\infty}^{+\infty} y_i^2 + \sum_{i=-\infty}^{+\infty} e^{x_i - x_{i+1}} + c \sum_{i=-\infty}^{+\infty} (x_i - x_{i+1}) \quad (1.17)$$

for some constant c.

Restricting to N particles $(x_i)_{i=1}^N$, and setting $c = 0$, one obtains the so-called *open Toda system* with Hamiltonian

$$H_{\text{Toda}}(x, y) = \frac{1}{2} \sum_{i=1}^{N} y_i^2 + \sum_{i=1}^{N-1} e^{x_i - x_{i+1}}. \quad (1.18)$$

In this monograph, we will, unless explicitly stated otherwise, restrict our attention to the open Toda system, which we call simply the Toda system (see Section 1.5).

Flaschka (1974a,b) and (independently) Manakov (1974) showed that Hamilton's equations

$$\partial_t x = \partial_y H_{\text{Toda}}, \qquad \partial_t y = -\partial_x H_{\text{Toda}}$$

can be written in Lax pair form (1.1), where $X = X(t)$ is tridiagonal,

$$\begin{cases} X_{ii} = -\frac{y_i}{2}, & 1 \leq i \leq N, \\ X_{i,i+1} = X_{i+1,i} = \frac{1}{2} e^{\frac{1}{2}(x_i - x_{i+1})}, & 1 \leq i \leq N-1 \end{cases} \quad (1.19)$$

and $B(X) = X_- - X_-^T$ is a (tridiagonal) skew-symmetric matrix as above. As noted by Flaschka, not only is the flow $t \mapsto X(t)$ isospectral, so that the eigenvalues $\lambda_i(t) = \lambda_i(0)$ give N constants of motion for the Toda flow, but they are independent and Poisson-commute in the underlying symplectic structure

$$\left(\mathbb{R}^{2n}, \omega = \sum_{i=1}^{N} dx_i \wedge dy_i \right).$$

---

[3] More detailed descriptions of this history can be found, for example, in Deift et al. (2022) and Kodama and Shipman (2018).

1.2 Description of the Algorithms

Thus, the Toda lattice is integrable in the sense of Liouville.[4]

In later work, Moser (1975a) showed how to solve the Toda lattice explicitly, and he also showed how to evaluate the long time behavior of the system. The Toda system (1.1) is the natural extension of the original tridiagonal Toda lattice to full $N \times N$ matrices. Although it is not a priori clear, this extended system is also Hamiltonian, and integrable in the sense of Liouville, and super-integrable as defined in Chapter 4. We will derive, and explain, all the above properties of the Toda system in Chapters 3 and 4.

We note here that it is a basic observation in the modern theory of integrable systems, that if a dynamical system can be written in Lax pair form

$$\partial_t S = [S, U] = SU - US$$

for some $U = U(S)$, then the flow $t \mapsto S(t)$ is isospectral, so that the system has at least N integrals of the motion, where $N = \dim S$. This follows in general by the argument – due to Lax – following (1.8).

1.2.2 The QR Algorithm

The QR algorithm, introduced by Francis in 1961, based on, but superseding the Rutishauser LR algorithm, works in the following way. Let X_0 be an invertible matrix in Σ_N. Then X_0 has a *QR factorization*

$$X_0 = Q_0 R_0, \tag{1.20}$$

where Q_0 is orthogonal and R_0 is upper-triangular with $(R_0)_{ii} > 0, i = 1 \ldots N$. The factorization is unique. The factorization is obtained by applying the Gram-Schmidt process to the columns of X_0 starting from the left. Set

$$X_1 = R_0 Q_0.$$

Substituting $R_0 = Q_0^T X_0$ from (1.20), we see that

$$X_1 = Q_0^T X_0 Q_0. \tag{1.21}$$

The *QR step* is the map φ_{QR} that switches Q and R,

$$X = QR \mapsto \varphi_{QR}(X) = RQ.$$

---

[4] The notions of *symplectic manifold*, *Hamiltonian flow*, *Poisson-commutativity*, as well as various notions of *integrability*, will be defined and discussed in Chapter 2.

From (1.21), we see that φ_{QR} is isospectral. Now X_1 in turn also has its own QR factorization,

$$X_1 = Q_1 R_1.$$

Set

$$X_2 = R_1 Q_1 = \varphi_{QR}(X_1)$$

and so on. In this way we obtain an isospectral sequence (X_k) of matrices. Generically, X_k converges as $k \to \infty$, to a diagonal matrix $X_\infty = \mathrm{diag}(\lambda_1, \ldots \lambda_N)$. Again, necessarily, the λ_i are the eigenvalues of X_0. This construction is at the heart of the so-called QR algorithm that plays an outsize role in numerical analysis, and occupies a prime position in software (see LINPACK) for eigenvalue computation. It turns out that there is a flow (the **QR flow**) given by a Lax pair

$$\partial_t X = [X, B(\ln X)], \qquad X(0) = X_0, \qquad (1.22)$$

with the property that at integer times it interpolates the QR steps starting with a positive definite matrix X_0. Moreover, in the case that X_0 is positive definite, this flow is Hamiltonian on an appropriate symplectic manifold (see Symes, 1980, 1982, Deift et al., 1983 and Chapter 3). Also, the flow is integrable. More precisely, we have the following result.

Theorem 1.2 *Let Σ'_N be the set of real symmetric positive definite matrices.*

(1) *(Stroboscope property) For $X \in \Sigma'_N$, set*

$$H_{QR}(X) := -\mathrm{tr}(X \ln X - X). \qquad (1.23)$$

Then H_{QR} generates the flow (1.22) with solution $X_{QR}(t)$, $X_{QR}(0) = X_0$, and, at integer times k,

$$X_{QR}(k) = \varphi_{QR}^k(X_0). \qquad (1.24)$$

(2) *(Integrability) The Hamiltonian H_{QR} is completely integrable on a symplectic manifold of real, symmetric matrices. Moreover, the flow $X_{QR}(t)$ commutes with the Toda flow generated by (1.1).*[5]

We will prove and explain the above properties of the QR algorithm in Chapter 3. We will also show that if $X \in \Sigma_N$ is invertible, but not positive definite, then H_{QR} generates a suitably modified flow on self-adjoint matrices.

---

[5] For more information on the symplectic structure for the H_{QR}-flow, see Section 3.8.

1.3 Goal of This Work

Our main goal is to prove the result of Deift and Trogdon (2018) on universality for the Toda eigenvalue algorithm to compute the top eigenvalue of a random real symmetric or Hermitian matrix H chosen from some random matrix ensemble. We concentrate first on the real symmetric case: The modifications that are necessary in the Hermitian case are given in Section 3.10.

The result is the following. Let $(X(t))_{t \geq 0}$ be the solution of (1.1) with $X(0) = H \in \Sigma_N$. Let

$$E(t) = \sum_{n=2}^{N} (X_{1n}(t))^2 \tag{1.25}$$

so that $E(t) = 0$ implies that $X_{11}(t)$ is an eigenvalue of H. Thus, the *halting time* (or the 1-*deflation time*) for the Toda algorithm is given by

$$T^{(1)}(H) = \inf\{t \geq 0 \, : \, E(t) < \varepsilon^2\}.$$

Note that by the min-max principle, if $E(t) < \varepsilon^2$, then $|X_{11}(t) - \lambda_j| < \varepsilon$ for some eigenvalue λ_j of $X(0) = H$. Moreover, the Toda algorithm is generically sorting, in the sense that the eigenvalues appear ordered in the limit:

$$X(t) \xrightarrow[t \to \infty]{} X_\infty = \mathrm{diag}(\lambda_1, \dots, \lambda_N), \quad \lambda_1 \geq \cdots \geq \lambda_N.$$

It follows that for generic $X(0) = H$, λ_1 above is the top eigenvalue of H.

For invariant and Wigner random matrix ensembles (see Section 5.1 for appropriate definitions) there is a constant $c_V > 0$, which depends on the ensemble, such that the following limit exists

$$F^{\mathrm{gap}}(t) = \lim_{N \to \infty} \mathbb{P}\left(\frac{1}{c_V^{2/3} 2^{-2/3} N^{2/3} (\lambda_1 - \lambda_2)} \leq t\right), \quad t \geq 0.$$

This limit depends only on the symmetry class (i.e., real symmetric versus complex Hermitian) of the random matrix ensemble under consideration. The basic universality result in Deift and Trogdon (2017) is the following.

Theorem 1.3 (See also Theorem 6.2) *Let $0 < \sigma < 1$ be fixed and let (ε, N) be in the scaling region*

$$\frac{\ln \varepsilon^{-1}}{\ln N} \geq \frac{5}{3} + \frac{\sigma}{2}. \tag{1.26}$$

Then, if $H \in \Sigma_N$ is distributed according to any invariant or Wigner ensemble, we have

$$\lim_{N \to \infty} \mathbb{P}\left(\frac{T^{(1)}}{c_V^{2/3} 2^{-2/3} N^{2/3} (\ln(\varepsilon^{-1}) - 2/3 \ln N)} \leq t\right) = F^{\mathrm{gap}}(t), \quad t \geq 0.$$

Thus, the halting time for the Toda algorithm to compute the top eigenvalue of a random matrix is universal, and behaves statistically like the inverse of the gap $\lambda_1 - \lambda_2$ between the two top eigenvalues of the random matrix. We emphasize that this not only gives an upper bound for the runtime of the algorithm but it describes the *true* behavior of the algorithm accurately as shown in Figures 6.1 and 6.2. We will prove this result, and more, including the case where H is Hermitian, in Chapter 6.

Remark 1.4 In the case of invariant ensembles, we will only prove Theorem 1.3 for weights that are exponentials of convex functions as in Definition 5.2. This is because the very detailed random matrix estimates required to prove Theorem 1.3 (see Chapters 5 and 6) are available in the literature, as of 2025, only in the convex case. The technology to prove the required results for a very general class of non-convex functions is available, and the requisite estimates just need to be established. Specifically, local properties of the equilibrium measure near the top eigenvalue, extending Theorem 5.3 to the non-convex case, are needed. Extensive numerical calculations certainly indicate that Theorem 1.3 indeed remains true, with great precision, in the non-convex case.

Note finally that the scaling regime (1.26), in which F^{gap} random matrix behavior is guaranteed to appear, includes a common arena for numerical computation. Indeed, for $\varepsilon = 10^{-16}$ and $N < 10^9$, we have

$$\frac{\ln \varepsilon^{-1}}{\ln N} > \frac{16}{9} > \frac{5}{3}.$$

1.4 Outline for the Monograph

The outline for this monograph is as follows:

- basics of Hamiltonian mechanics and integrable systems (Chapter 2);
- properties of the Toda lattice and its generalizations (Chapter 3);
- an alternative study of the Toda algorithm without relying on a Hamiltonian structure (Chapter 4);
- properties of random matrix ensembles (Chapter 5); and
- the proof of universality for the Toda algorithm to compute the top eigenvalue of a random matrix (Chapter 6).

In these notes, we will present far more properties of the Toda lattice than are needed to prove Theorem 1.3. In fact, all that is needed are the results from Section 3.4. Our broader goal, however, is to showcase for the interested reader some of the remarkable properties of the Toda lattice, some of which are new and have never been written down before.

1.5 Related Scenarios and Some Notational Issues

In this monograph we are concerned only with the open Toda lattice and its natural extensions. However, the Toda flow generated by (1.17) has been analyzed in great detail in a variety of different scenarios.

Here are some examples:

1. Prior to the analysis of the open Toda lattice, Flaschka (1974a,b) and Manakov (1974) considered the flow generated by (1.17) in the case $c = 0$ subject to periodic boundary conditions (the so-called periodic Toda lattice),

$$x_{n+N} = x_n, \; y_{n+N} = y_n$$

   for some $N < \infty$. In 1976, Date and Tanaka, expanding on earlier partial results of Kac and Van Moerbeke (1975), showed how to solve the periodic Toda lattice explicitly in terms of hyperelliptic functions.

2. Flaschka (1974a) and Manakov (1974) also analyzed the flow generated by (1.17) in the case $c = 0$ with an infinite number of particles subject to

$$x_n - x_{n+1} \to d = constant, \; y_n \to 0$$

   as $n \to \pm\infty$, using inverse scattering methods.

3. In 1991, Venakides et al. analyzed the flow generated by (1.17), again in the case $c = 0$, with initial conditions

$$x_n(t = 0) = n\delta, \; y_n(t = 0) = 0, \; n = 1, 2, \ldots, \infty$$

   for some fixed $\delta > 0$, and with $x_0(t) = vt$ for all $t \geq 0$ for some $v > 0$ (the so-called Toda shock problem). Here the authors analyzed the flow using inverse scattering and the Lax–Levermore–Venakides method.

4. In 1975, considered the same problem as (3), but now with $v < 0$ (the so-called Toda rarefaction problem). Here the authors used the Riemann–Hilbert/steepest descent method.

5. Over the last 3–4 years, Egorova et al. (2023) have analyzed the flow generated by (1.17), again in the case $c = 0$, with an infinite number of particles subject to

$$x_n - x_{n+1} \to d_\pm, \; y_n \to e_\pm$$

as $n \to \pm\infty$. Here $d_+ \neq d_-$ and $e_+ \neq e_-$ and the authors use the Riemann–Hilbert/steepest-descent method.

6. In 2022, Deift, Li, Spohn, Tomei and Trogdon considered the flow generated by (1.17) with a finite number N of particles and $c \neq 0$ (the so-called open Toda chain with external forcing). In the case $c > 0$ the external force stretches the chain, and for $c < 0$ the force compresses the chain. The authors show that for $c > 0$, the chain remains integrable. For $c<0$, the problem remains open.

7. There are counterparts to Toda flows and QR algorithms related to the singular values of matrices. In 1986, Chu introduced the flow for singular values and Li (1986) proved integrability for the flow on generic orbits.

8. In 2023, Leite, Saldanha, Tomei and Torres converted the Toda and SVD flows into straight line motion in Euclidean space using the Z-variables defined in Chapter 4. This represents a radically new and fruitful approach to the Toda lattice, as explained in Chapter 3.

The above list of examples is very far from complete and much seminal work is not represented. We ask the readers for their indulgence on this score: The list is intended only to give them some exposure to the scope of Toda theory over the years to the present time.

Regarding notation used in this monograph, we note the following:

- A Jacobi matrix is a real symmetric tridiagonal matrix with positive off diagonal entries (see Definition 3.4). Most authors denote the diagonal entries in a Jacobi matrix by b_i and the off-diagonal entries by a_i. In this monograph we reverse the role of the a_i and b_i as in (3.4), for no other reason than that some of the present authors made this choice when they first began working on the Toda lattice in the early 1980s.

- From a spectral theory point of view, it is usual to write the spectral theorem for a real symmetric $n \times n$ matrix A as $A = Q\Lambda Q^T$, where the columns q_i of Q are the orthonormalized eigenvectors of A and $\Lambda = \text{diag}(\lambda_1, \ldots, \lambda_n)$ are the corresponding eigenvalues, $Aq_i = \lambda_i q_i$. In this monograph we often write $A = Q^T \Lambda Q$ so that the rows of Q are now the eigenvectors of A. This representation arises naturally in the solution of the Toda lattice as in Theorem 1.1, and also in the analysis of the Toda lattice without Hamiltonian structure in Chapter 4.

- The Toda equations are usually written in Lax pair form,

$$\partial_t X = [X, B(X)],$$

1.5 Related Scenarios and Some Notational Issues

where $B(X) = X_- - X_-^T$, as in (1.1). In discussing the Lie–Poisson structure of the Toda equations, however, it is natural to write the equations in the form $\partial_t X = [X, \pi_k(X)]$, where $\pi_k(X) = X_+ - X_+^T$, as in (3.106) in the case with $q = 2$. However, for a symmetric matrix X, $B(X) = -\pi_k X$, so the solution of (3.106) at time t is just the solution of (1.1) at time $-t$.

In numerical linear algebra, in addition to the factorization $X = QR$, one often uses the factorization $X = \hat{Q}L$, where $\hat{Q}$ is again orthogonal, but L is now lower triangular with positive diagonal entries: This factorization is just the Gram–Schmidt process applied to the columns of X, but now starting from the right. The relationship between $B(X)$ and $\pi_k(X)$ reflects the relationship between the QR and $\hat{Q}L$ factorizations. Indeed, as noted in Theorem 3.32 with $q = 2$, if X is real symmetric and $e^{tX} = Q(t)R(t)$, then $X(t) = Q(t)^T X Q(t)$ is the solution of (1.1). On the other hand, if

$$e^{tX} = \hat{Q}(t)L(t),$$

then a similar computation to that in Theorem 3.32 shows that

$$\hat{X}(t) = \hat{Q}(t)^T X \hat{Q}(t)$$

solves (3.106) with $q = 2$. But from $e^{tX} = \hat{Q}(t)L(t)$, we obtain

$$e^{-tX} = \hat{Q}(t)(L(t)^T)^{-1}.$$

However, $e^{-tX} = Q(-t)R(-t)$, and by the uniqueness of the QR factorization, we see that $\hat{Q}(t) = Q(-t)$, and so

$$X(-t) = Q(-t)^T X Q(-t) = \hat{Q}(t)^T X \hat{Q}(t) = \hat{X}(t),$$

as noted above.

2
Hamiltonian Mechanics and Integrable Systems

The goal of this chapter is to give an elementary introduction to Hamiltonian mechanics, and particularly integrable Hamiltonian systems, with a view to describing various results that we need in analyzing the Toda algorithm. The reader is encouraged to consult references such as Abraham and Marsden (1978), Arnold (1978), Kirillov (2004), Moser and Zehnder (2005) and Warner (1983) for a more detailed presentation.

2.1 Hamiltonians on Symplectic Manifolds

We say that a space M is a *symplectic manifold*, if

- it is an even-dimensional manifold,
- with a nondegenerate 2-form ω, i.e.,[1] for any $u = T_m(M)$,

$$\omega(u,v) = 0 \quad \forall v \in T_m(M) \quad \text{implies} \quad u = 0,$$

- which is closed, i.e., $d\omega = 0$.

Note that non-degeneracy and skew-symmetry ($\omega(u,v) = -\omega(v,u)$) imply that M has even dimension; we will sometimes write $M^{(2n)}$.

The space $M = \mathbb{R}^{2n}$ can be endowed with the structure of a symplectic manifold: The standard 2-form is $\omega = \sum_{i=1}^{N} dx_i \wedge dy_i$. Clearly, $d\omega = 0$. If

$$v = \sum_{i=1}^{N} a_i \partial_{x_i} + b_i \partial_{y_i} = \begin{pmatrix} a \\ b \end{pmatrix} \quad \text{and} \quad v' = \begin{pmatrix} a' \\ b' \end{pmatrix},$$

[1] Here $T_m(M) = T_m M$ denotes the tangent space at $m \in M$.

2.1 Hamiltonians on Symplectic Manifolds

then a simple calculation shows that

$$\omega(v,v') = \left\langle \begin{pmatrix} a \\ b \end{pmatrix}, J\begin{pmatrix} a' \\ b' \end{pmatrix} \right\rangle, \text{ where } J = \begin{pmatrix} 0 & I \\ -I & 0 \end{pmatrix},$$

where I is the $N \times N$ identity and $\langle \cdot, \cdot \rangle$ denotes the standard inner product on $\mathbb{R}^{2n}$. From this, it is clear that

$$\forall v', \quad \omega(v,v') = 0 \quad \Rightarrow \quad J^T\begin{pmatrix} a \\ b \end{pmatrix} = 0 \quad \Rightarrow \quad \begin{pmatrix} a \\ b \end{pmatrix} = 0,$$

i.e., $v = 0$, so ω is indeed nondegenerate.

Functions (or "Hamiltonians") $H : M \to \mathbb{R}$ generate vector fields through ω in the following way. At any point $m \in M$, $\mathrm{d}H_m$ is a 1-form. Hence,

$$v \mapsto \mathrm{d}H_m(v)$$

is a linear map from $T_m(M)$ to $\mathbb{R}$. Thus, as ω is nondegenerate, an extension of the usual Riesz representation theorem (on finite-dimensional vector spaces) implies the existence a unique vector $v_H(m) \in T_m(M)$ such that

$$\mathrm{d}H(v) = \omega(v_H, v) \text{ for all } v \in T_m M. \tag{2.1}$$

In the case of $\left(\mathbb{R}^{2n}, \omega = \sum_{i=1}^N \mathrm{d}x_i \wedge \mathrm{d}y_i\right)$, equation (2.1) becomes for $v = \begin{pmatrix} a \\ b \end{pmatrix}$,

$$\mathrm{d}H(v) = \sum_{i=1}^N (a_i \partial_{x_i} H + b_i \partial_{y_i} H) = \left\langle \begin{pmatrix} H_x \\ H_y \end{pmatrix}, \begin{pmatrix} a \\ b \end{pmatrix} \right\rangle = \left\langle v_H, J\begin{pmatrix} a \\ b \end{pmatrix} \right\rangle.$$

Hence $\begin{pmatrix} H_x \\ H_y \end{pmatrix} = J^T v_H$, which implies as $JJ^T = I_{2n}$,

$$v_H = J\begin{pmatrix} H_x \\ H_y \end{pmatrix} = \begin{pmatrix} H_y \\ -H_x \end{pmatrix}.$$

Thus H gives rise to the standard Hamiltonian vector field

$$\partial_t \begin{pmatrix} x \\ y \end{pmatrix} = v_H = \begin{pmatrix} H_y \\ -H_x \end{pmatrix}.$$

The fact that ω is nondegenerate plays an obvious role here. The role of the assumption that ω is closed ($\mathrm{d}\omega = 0$) is more subtle. If H, K are two Hamiltonians on M, we define their **Poisson bracket** $\{H, K\}$ through

$$\{H, K\} := \omega(v_H, v_K). \tag{2.2}$$

Note that

$$\{H, K\} = -\{K, H\},$$

and that, as ω is nondegenerate, $\{H, K\} = 0$ for all K implies that H is a constant. Clearly, $\{H, K\} = dH(v_K) = v_K(H) = \partial_t H$, where $\partial_t H$ is the derivative of H in the direction of v_K. Thus,

$$\partial_t H = \{H, K\} \tag{2.3}$$

describes the change of H along the flow generated by K. In particular,

$$\partial_t K = \{K, K\} = 0,$$

as $\{\cdot, \cdot\}$ is skew. Thus K is a conserved quantity for the flow it generates. The Poisson bracket acts like a derivative and satisfies Leibniz' rule,

$$\{HK, L\} = H\{K, L\} + \{H, L\}K \tag{2.4}$$

for all Hamiltonians H, K, and L. Indeed, under the flow generated by L,

$$\{HK, L\} = \partial_t(HK) = \partial_t(H)K + H\partial_t(K) = \{H, L\}K + H\{K, L\}.$$

Observe that for $\left(\mathbb{R}^{2n}, \sum_{i=1}^{N} dx_i \wedge dy_i\right)$,

$$\{H, K\} = \omega(v_H, v_K) = \left\langle J\begin{pmatrix} H_x \\ H_y \end{pmatrix}, J^2\begin{pmatrix} K_x \\ K_y \end{pmatrix} \right\rangle = \left\langle \begin{pmatrix} H_x \\ H_y \end{pmatrix}, J\begin{pmatrix} K_x \\ K_y \end{pmatrix} \right\rangle,$$

i.e.,

$$\{H, K\} = \sum_{i=1}^{N} (\partial_{x_i} H \partial_{y_i} K - \partial_{y_i} H \partial_{x_i} K).$$

Thus

$$\{\{H, K\}, L\} = \sum_{i=1}^{N} (\partial_{x_i}\{H, K\}\partial_{y_i} L - \partial_{y_i}\{H, K\}\partial_{x_i} L)$$

$$= \sum_{i=1}^{N} \partial_{x_i}\left(\sum_{k=1}^{N}(\partial_{x_k} H \partial_{y_k} K - \partial_{y_k} H \partial_{x_k} K)\right)\partial_{y_i} L$$

$$- \partial_{y_i}\left(\sum_{k=1}^{N}(\partial_{x_k} H \partial_{y_k} K - \partial_{y_k} H \partial_{x_k} K)\right)\partial_{x_i} L,$$

which by cancellations due to symmetries, leads to the **Jacobi identity**:

$$\{\{H, K\}, L\} + \{\{K, L\}, H\} + \{\{L, H\}, K\} = 0. \tag{2.5}$$

2.1 Hamiltonians on Symplectic Manifolds

Now a standard calculation using the calculus of forms, shows that

$$d\omega(v_H, v_K, v_L) = C\left(\{\{H,K\},L\} + \{\{K,L\},H\} + \{\{L,H\},K\}\right) \quad (2.6)$$

for some constant C. Thus, H, K, L satisfy Jacobi's identity if and only if ω is closed, i.e., $d\omega = 0$. Jacobi's identity leads to the following critical calculation for the commutator $[v_H, v_K]$ of Hamiltonian vector fields:

$$\begin{aligned}
{[v_H, v_K](L)} &= v_H(v_K(L)) - v_K(v_H(L)) = v_H(\{L,K\}) - v_K(\{L,H\}) \\
&= \{\{L,K\},H\} - \{\{L,H\},K\} = \{\{L,K\},H\} + \{\{H,L\},K\} \\
&= -\{\{K,H\},L\} = \{L,\{K,H\}\} = v_{\{K,H\}}(L),
\end{aligned}$$

which is to say

$$[v_H, v_K] = v_{\{K,H\}}. \quad (2.7)$$

Thus, the commutator of two Hamiltonian vector fields is again a Hamiltonian vector field. Moreover, the map $H \mapsto v_H$ is an anti-isomorphism from the Poisson algebra of functions given by the Poisson bracket, into the algebra of vector fields with product given by the commutator of vector fields. Most importantly, we notice that two Hamiltonian vector fields commute if and only if their Hamiltonians Poisson-commute. This is perhaps the most useful consequence of the fact that ω is closed.

It is a general fact for vector fields on a manifold that $v, \tilde{v}$ commute if and only if the flows $\phi_v^t, \phi_{\tilde{v}}^t$ that they generate commute, i.e., if

$$\partial_t \phi_v^t = v(\phi_v^t) \quad \text{and} \quad \partial_t \phi_{\tilde{v}}^t = \tilde{v}(\phi_{\tilde{v}}^t),$$

then for every s, t for which the flows exists,

$$\phi_v^t \circ \phi_{\tilde{v}}^s = \phi_{\tilde{v}}^s \circ \phi_v^t. \quad (2.8)$$

In particular, flows generated by Hamiltonians commute if and only if their Hamiltonians Poisson-commute.

From the preceding calculations, we see that symplectic manifolds give rise to nondegenerate *Poisson manifolds* $(M, \{\cdot, \cdot\})$ where the Poisson bracket

- is bilinear from $\mathcal{C}^\infty(M) \times \mathcal{C}^\infty(M) \mapsto \mathcal{C}^\infty(M)$;
- satisfies Leibniz' rule: $\{HK, L\} = H\{K,L\} + \{H,L\}K$;
- is nondegenerate, i.e., $\{H,K\} = 0$ for all functions K implies that H is a constant; and
- satisfies the Jacobi identity (2.5).

Exercise 2.1 Prove the converse, i.e., if $(M, \{\cdot, \cdot\})$ is a nondegenerate Poisson manifold with the four above properties, then (M, ω) is a symplectic manifold where

$$\omega(v_H, v_K) := \{H, K\}$$

and

$$v_H(L) = \{L, H\}, \quad v_K(L) = \{L, K\}.$$

2.2 How Symplectic Manifolds Arise

There are, amongst others, three natural sources of symplectic manifolds:

- T^*X, i.e., cotangent bundles on manifolds,
- co-adjoint orbits of groups on their dual Lie algebras,[2] and
- constrained systems.

We now say a few words about each of these situations.

(1) Cotangent bundles

Let X be a manifold, $x \in X$, and let $\alpha_x \in T_x^*X$. Let v be a vector field on

$$T^*X = \bigcup_{x \in X} T_x^*X$$

and let π denote the natural projection on T^*X to the base point x. Thus, $\pi(\alpha_x) = x$ and $\pi_*(v)$, the push-forward of v under π, lies in $T_x X$, as represented on Figure 2.1.

Hence, as $\alpha_x \in T_x^*X$,

$$\theta(\alpha_x)(v) := -\alpha_x(\pi_*(v))$$

defines a natural 1-form θ on T^*X. Then, $\omega = d\theta$ defines a 2-form on T^*X, which is clearly closed (as $d\omega = d^2\theta = 0$) and can easily be shown to be nondegenerate.

Exercise 2.2 On $\mathbb{R}^{2n} = T^*\mathbb{R}^n$, show that

$$\theta = -\sum y_i dx_i$$

[2] In that case the 2-form is called the Kostant–Kirillov form.

2.2 How Symplectic Manifolds Arise

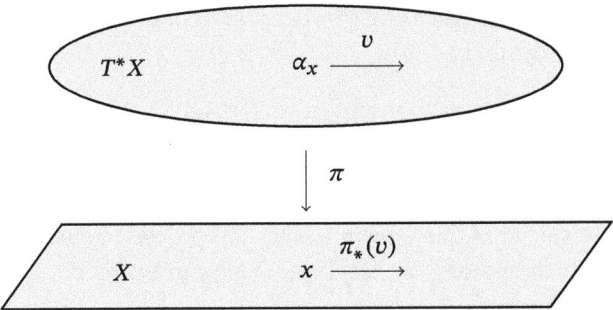

Figure 2.1 The cotangent bundle T^*X over the manifold X is a manifold, canonically endowed with the projection π.

and hence

$$\omega = d\theta = \sum_{i=1}^{N} dx_i \wedge dy_i.$$

The above construction leads to the appearance of symplectic manifolds in many different situations; we refer the interested reader to Warner (1983).

(2) Orbits in Lie algebras

Let $\mathfrak{g}$ be a Lie algebra with bracket $[\cdot,\cdot]$ that

- is bilinear and
- satisfies the Jacobi identity, i.e.,

$$[[x,y],z] + [[y,z],x] + [[z,x],y] = 0 \quad \forall x,y,z \in \mathfrak{g}.$$

Let G be the associated connected group with identity e_G and let $\mathfrak{g}^*$ be the dual Lie algebra. Recall that $\mathfrak{g} \simeq T_{e_G}G$. Then G acts on $\mathfrak{g}$ by the Ad-action:

$$\mathrm{Ad} : \mathfrak{g} \to \mathfrak{g}$$
$$x \mapsto \partial_t|_{t=0}\, g e^{tx} g^{-1}, \quad g \in G, x \in \mathfrak{g}.$$

Here, $t \mapsto e^{tx}$, $e^{tx}|_{t=0} = e_G$ is the unique 1-parameter subgroup of G whose tangent vector at e_G is x. Also, G acts on $\mathfrak{g}^*$ by adjointness:

$$\mathrm{Ad}^* : \mathfrak{g}^* \to \mathfrak{g}^*$$
$$\langle \mathrm{Ad}_g^* \alpha, x \rangle = \langle \alpha, \mathrm{Ad}_g x \rangle = \alpha(\mathrm{Ad}_g x),$$

where $\langle \cdot, \cdot \rangle$ denotes the pairing of $\mathfrak{g}^*$ with $\mathfrak{g}$. The co-adjoint orbit $\mathcal{O}_\alpha$ through a point $\alpha \in \mathfrak{g}^*$ is given by

$$\mathcal{O}_\alpha = \{\mathrm{Ad}_g^* \alpha \ : \ g \in G\}. \tag{2.9}$$

The remarkable fact is that $\mathcal{O}_\alpha$ carries a nondegenerate closed 2-form and hence is naturally a symplectic manifold (and hence an even-dimensional manifold). To see what this 2-form is, one can proceed functorially. Vector fields on $\mathcal{O}_\alpha$ at the point $\beta = \mathrm{Ad}_g^* \alpha \in \mathcal{O}_\alpha$ are given by $\partial_t|_{t=0} \mathrm{Ad}_{e^{tx}}^* \beta$ for arbitrary $g \in \mathfrak{g}$. Indeed,

$$\partial_t|_{t=0} \mathrm{Ad}_{e^{tx}}^* \beta = \partial_t|_{t=0} \mathrm{Ad}_{e^{ty}}^* \mathrm{Ad}_g^* \alpha = \partial_t|_{t=0} \mathrm{Ad}_{ge^{tx}}^* \alpha \in T_\alpha \mathcal{O}_\alpha.$$

It is natural to define

$$\omega_\beta \left(\partial_t|_{t=0} \mathrm{Ad}_{e^{tx}}^* \beta, \partial_t|_{t=0} \mathrm{Ad}_{e^{ty}}^* \beta \right) := \beta([x,y]) = \langle \beta, [x,y] \rangle$$

and indeed, with this definition, $\mathcal{O}_\alpha$ is a symplectic manifold.

Example 2.3

$$G = \mathrm{GL}_N^+(\mathbb{R}) = \mathrm{GL}_N(\mathbb{R}) \cap \{M \ : \ \det M > 0\},$$
$$\mathfrak{g} = \mathfrak{gl}_N(\mathbb{R}) = M_N(\mathbb{R}).$$

We can identify $\mathfrak{g}^*$ as $M_N(\mathbb{R})$ through the nondegenerate pairing

$$\langle A, B \rangle = \mathrm{tr}(AB), \tag{2.10}$$

i.e., the matrix A induces a linear form l on $\mathfrak{g}$ through $B \mapsto l(B) = \mathrm{tr}(AB)$. Every linear map l is of this form for some unique $A = A(l)$. Now for $g \in G, x \in \mathfrak{g}$,

$$\mathrm{Ad}_g x = \partial_t|_{t=0} g e^{tx} g^{-1} = g x g^{-1},$$

where the right-hand side can now be taken to be ordinary matrix multiplication. Also, for all $x \in \mathfrak{g} = M_N(\mathbb{R})$,

$$\langle \mathrm{Ad}_g^* A, x \rangle = \langle A, \mathrm{Ad}_g x \rangle = \langle A, gxg^{-1} \rangle = \mathrm{tr}(Agxg^{-1}) = \mathrm{tr}(g^{-1}Agx).$$

Thus, $\mathrm{Ad}_g^* A = g^{-1} A g$ and hence

$$\mathcal{O}_A = \{g^{-1} A g \ : \ g \in G\},$$

the set of all real matrices that are $\mathrm{GL}_N^+(\mathbb{R})$-conjugated to A.

Exercise 2.4 Compute all possible co-adjoint orbits $\mathcal{O}_A$ for 2×2 matrices.

2.2 How Symplectic Manifolds Arise

Each $x \in \mathfrak{g}$ induces a function H_x on $\mathfrak{g}^*$ in a natural way as follows:

$$H_x(A) = \langle A, x \rangle = \mathrm{tr}(Ax), \qquad A \in \mathfrak{g}^*. \tag{2.11}$$

For such functions H_x, we have

$$\begin{aligned} \mathrm{d}H_x\left(\partial_t|_{t=0}\, \mathrm{Ad}^*_{e^{ty}} A\right) &= \partial_t|_{t=0} H_x(\mathrm{Ad}^*_{e^{ty}} A) = \partial_t|_{t=0} H_x(e^{-ty} A e^{ty}) \\ &= \partial_t|_{t=0} \mathrm{tr}(e^{-ty} A e^{ty} x) = \mathrm{tr}(A[y,x]) = \langle A, [y,x] \rangle \\ &= \omega\left(\partial_t|_{t=0}\, \mathrm{Ad}^*_{e^{ty}} A, \partial_t|_{t=0}\, \mathrm{Ad}^*_{e^{tx}} A\right). \end{aligned}$$

Thus,

$$v_{H_x}(A) = -\partial_t|_{t=0} \mathrm{Ad}^*_{e^{tx}} A = -\partial_t|_{t=0} e^{-tx} A e^{tx} = [x, A] \tag{2.12}$$

and

$$\{H_x, H_y\} = \omega_A(v_{H_x}, v_{H_y}) = \mathrm{tr}(A[x,y]). \tag{2.13}$$

The differential $\mathrm{d}H_x(A)$ is by definition a linear functional on $\mathfrak{g}^*$, and hence the (unique) element in $(\mathfrak{g}^*)^* = \mathfrak{g}$ such that

$$\mathrm{d}H_x(A)(B) = \partial_t|_{t=0} H_x(A + tB) = \partial_t|_{t=0} \mathrm{tr}\left(x(A + tB)\right) = \mathrm{tr}(xB). \tag{2.14}$$

Thus, $\mathrm{d}H_x(A) = x$ and we see that

$$\{H_x, H_y\}(A) = \mathrm{tr}\left(A[\mathrm{d}H_x(A), \mathrm{d}H_y(A)]\right). \tag{2.15}$$

By linearity and Leibniz' rule, we conclude, using Exercise 2.5, that

$$\{H, K\}(A) = \mathrm{tr}\left(A[\mathrm{d}H(A), \mathrm{d}K(A)]\right) \tag{2.16}$$

for arbitrary smooth functions H and K on $\mathfrak{g}^*$.

Exercise 2.5 Show that arbitrary smooth functions on $\mathfrak{g}^*$ can be approximated by finite linear combinations of sums of powers of functions of type H_x.

On a general dual Lie algebra $\mathfrak{g}^*$, (2.16) becomes

$$\{H, K\}(\alpha) = \langle \alpha, [\mathrm{d}H(\alpha), \mathrm{d}K(\alpha)] \rangle, \qquad \alpha \in \mathfrak{g}^*, \tag{2.17}$$

where $\langle x, y \rangle$ denotes the action of $\mathfrak{g}^*$ on $\mathfrak{g}$, $x \in \mathfrak{g}^*$ and $y \in \mathfrak{g}$. In the case of $\mathrm{GL}_n^+(\mathbb{R})$, for $H : M_N(\mathbb{R}) \to \mathbb{R}$, we have

$$\partial_t|_{t=0} H(A + tB) = \sum_{ij} (\partial_{A_{ij}} H) B_{ij} = \mathrm{tr}(\nabla H^T(A) B), \tag{2.18}$$

where $\nabla H(A)$ is the matrix with entries $\left(\partial_{A_{ij}} H\right)$. Thus,

$$\{H, K\}(A) = \mathrm{tr}\left(A[\nabla H^T(A), \nabla K^T(A)]\right). \tag{2.19}$$

We have constructed a Poisson manifold, i.e., $\mathfrak{g}^*$ is a manifold with a bracket $\{\cdot,\cdot\}$ which satisfies all the conditions for a Poisson bracket (see (2.2) and below); in general, however, it is degenerate. By general theory (see, e.g., Abraham and Marsden, 1978) $\mathfrak{g}^*$ is foliated by symplectic leaves, i.e., submanifolds of $\mathfrak{g}^*$ which are symplectic. It turns out that these symplectic leaves are precisely the co-adjoint orbits $\mathcal{O}_\alpha$. We will say much more about these orbits $\mathcal{O}_\alpha$ in Section 3.4; in particular, we will be interested in the group L of lower triangular matrices with its Lie and dual Lie algebra, l and l^*, respectively.

Exercise 2.6 Compute the symplectic leaves for $\mathfrak{g}^*$ if $G = \mathrm{GL}_N^+(\mathbb{R})$.

Exercise 2.7 For $G = \mathrm{GL}_N^+(\mathbb{R})$ as above, verify directly that:

(1) $\mathcal{O}_\alpha$ is a manifold immersed in $\mathrm{GL}_N^+(\mathbb{R})$ for any α.
(2) Vectors of the form

$$(v_x)_\beta = \partial_t|_{t=0} \mathrm{Ad}^*_{e^{tx}} \beta = \partial_t|_{t=0} e^{-tx} \beta e^{tx} = [\beta, x], \quad \beta \in \mathcal{O}_\alpha,$$

span the tangent space to $\mathcal{O}_\alpha$ at β.
(3) $\omega_\beta\big((v_x)_\beta, (v_y)_\beta\big)$ defined as $\langle \beta, [x, y]\rangle = \mathrm{tr}(\beta[x,y])$ is indeed a closed nondegenerate 2-form on $\mathcal{O}_\alpha$, $\beta \in \mathcal{O}_\alpha$, so that $\mathcal{O}_\alpha$ is a symplectic manifold.
(4) Verify directly that the Poisson bracket $\{H, K\}$ given by (2.19) satisfies the Jacobi identity (2.5).

Exercise 2.8 If the base manifold M is a group, then the manifolds in (1) and (2) are related. Basically, the symplectic structure in (2) is the pullback of the structure in (1) to the identity in G. Make this explicit.

Remark 2.9 A Hamiltonian K on a Poisson manifold generates a flow via (2.3) $\partial_t H = \{H, K\}$ even though the Poisson bracket may be degenerate.

For additional information on Kostant–Kirillov forms, the reader may consult Kirillov (2004).

(3) Constrained systems

(See, e.g., Deift et al., 1980; Moser, 1981) Suppose we have n one-dimensional harmonic oscillators:

$$\ddot{x}_i + \lambda_i x_i = 0, \quad 1 \leq i \leq n. \tag{2.20}$$

These equations are generated by the Hamiltonian

$$H = \frac{1}{2}\sum_{i=1}^{N}(y_i^2 + \lambda_i x_i^2) \qquad (2.21)$$

on the symplectic manifold $\left(\mathbb{R}^{2n}, \sum_{i=1}^{N} dx_i \wedge dy_i\right)$. Suppose we now constrain the oscillators to lie on the sphere, defined by

$$\phi_1 = 0, \qquad \phi_1(x) := \sum_{i=1}^{N} x_i^2 - 1. \qquad (2.22)$$

How would we describe the motion? Recalling from our first physics course how we solved the problem of a particle moving along a wire using Lagrangian mechanics, we could proceed accordingly by minimizing the constrained Lagrangian. The Hamiltonian version of this procedure, perhaps less familiar, is the following. Let

$$\phi_2(x,y) = \sum_{i=1}^{N} x_i y_i$$

and

$$X = \{(x,y) \in \mathbb{R}^{2n} : \phi_1 = \phi_2 = 0\}.$$

Clearly, the constrained motion should lie on $X \subset \mathbb{R}^{2n}$, as $y_i = \partial_t x_i$ under the flow generated by H and

$$\phi_2 = \sum_{i=1}^{N} x_i \partial_t x_i = \frac{1}{2}\partial_t \phi_1 = 0.$$

Moreover, X is even-dimensional and carries a natural 2-form $i^*\omega$, the pullback of ω under the immersion $i : X \to \mathbb{R}^{2n}$,

$$i^*\omega(u,v) = \omega(i_*u, i_*v),$$

where i_*u, i_*v are the push-forward of u, v from $T_x X \to T_{i(x)}\mathbb{R}^{2n}$. Recall that if f is a smooth function on $\mathbb{R}^{2n}$, then $f \circ i$ is a smooth function on X and

$$i_*v(f) = v(f \circ i).$$

Alternatively, $i^*\omega$ is just the restriction of ω to $TX \subset T\mathbb{R}^{2n}$. As the operator d commutes with the restriction (or the pull-back) operation, it is clear that $d(i^*\omega) = 0$, i.e., $i^*\omega$ is closed. The only question is whether it is nondegenerate.

Exercise 2.10 Show that $\{\phi_1, \phi_2\} \neq 0$, and conclude that $i_*\omega$ is nondegenerate.

Exercise 2.11 Compute the equations of motion for the above constrained flow. This constrained system is called the Neumann system and is in fact an integrable system (see Moser, 1981). It turns out (see Deift et al., 1980) that many well-known integrable systems such as the nonlinear Schrödinger equation or the Sine–Gordon equation can be obtained by constraining independent harmonic oscillators to quadrics, as in (2.22). More precisely, constraining a system of n one-dimensional oscillators to a finite number of quadrics as in (2.22), one obtains an integrable Hamiltonian system which converges as $n \to \infty$ to a partial differential equation. One particular choice of quadrics leads to the nonlinear Schrödinger equation: Another choice leads to the Sine–Gordon equation, and so on. In other words, such well-known integrable PDEs are just constrained harmonic oscillators, in disguise.

Suppose (M_1, ω_1) and (M_2, ω_2) are two symplectic manifolds and suppose that ψ is a map from $M_1 \to M_2$. We say that ψ is **symplectic** if $\psi^* \omega_2 = \omega_1$, i.e., ω_1 is the pullback of ω_2 under ψ. If and $(M_1, \{\cdot, \cdot\}_1)$ and $(M_2, \{\cdot, \cdot\}_2)$ are the Poisson manifolds arising as above from (M_1, ω_1) and (M_2, ω_2) respectively, then $\psi : M_1 \to M_2$ is symplectic if and only if

$$\{K \circ \psi, L \circ \psi\}_{(M_1, \{\cdot, \cdot\}_1)} = \{K, L\}_{(M_2, \{\cdot, \cdot\}_2)}$$

for all smooth K, L.

It is a fundamental fact that if ϕ^t is the flow generated by a Hamiltonian H on a symplectic manifold M,

$$\partial_t \phi^t = \{\phi^t, H\}$$

then for all t, the map $\phi^t : M \to M$ is symplectic.

2.3 Integration of Vector Fields

How can we integrate a dynamical system

$$\partial_t x(t) = V(x(t)), \quad x(t=0) = x_0, \qquad (2.23)$$

in $\mathbb{R}^m$? Suppose we have $m-1$ independent conserved quantities $\phi_1, \ldots, \phi_{m-1}$,

$$\partial_t \phi_j(x(t)) = 0, \quad 1 \le j \le m-1,$$

for solutions $x(t)$ of (2.23). Then we could in principle solve for $(m-1)$ variables in favor of the first one, and then we would be left with the equation

$$\partial_t x_1 = V(x_1, x_2(x_1), \ldots, x_m(x_1)),$$

which can be integrated by quadrature

$$\int_{x_1(0)}^{x_1(t)} \frac{du}{V(u, x_2(u), \ldots, x_m(u))} = t. \qquad (2.24)$$

(It is a remarkable fact that the Toda flow indeed has such a full set of integrals (see Corollary 4.13).)

In the theory of Hamiltonian systems, as opposed to general systems, a remarkable reduction occurs. One can solve systems of dimension $m = 2n$ which have only n independent integrals $\phi_1, \ldots, \phi_N$, provided that those integrals have the additional property

$$\{\phi_i, \phi_j\} = 0 \quad \text{for} \quad 1 \leq i, j \leq n, \qquad (2.25)$$

i.e., if the integrals Poisson-commute. As the Hamiltonian H for the system is conserved, we may always take one of the integrals, say $\phi_1 = H$. Note that as ϕ_i are integrals of the flow generated by H,

$$\{\phi_j, H\} = \partial_t \phi_j = 0 \qquad (2.26)$$

so the remaining integrals $\phi_2, \ldots, \phi_N$ Poisson-commute with $\phi_1 = H$.

A central result in the subject is the Liouville–Arnold–Jost (LAJ) theorem. We need some notation: A Hamiltonian vector field v_H on a symplectic manifold $(M^{(2n)}, \omega)$ is called **integrable (or completely integrable) in the sense of LAJ** on a domain $D \subset M^{(2n)}$ if it possesses n integrals

$$\phi_1 = H, \phi_2, \ldots, \phi_N,$$

which are linearly independent on D (i.e., $d\phi_1 \ldots d\phi_N$ are linearly independent at all points of D) and which Poisson-commute. We require that D is invariant under the flow generated by H for all t. This is a critical requirement as any Hamiltonian system is locally integrable, i.e., given any H and a point $m \in M^{(2N)}$ there exists a neighborhood B of m such that H has n independent Poisson-commuting integrals in this neighborhood. In general, the flow generated by H escapes from B and so these local integrals teach us nothing about the global behavior of the flow. The invariance of D is essential.

Exercise 2.12 Prove that any Hamiltonian system is locally integrable.

Theorem 2.13 (LAJ) *Suppose that $H = \phi_1$ is integrable on a domain $D \subset M = M^{2N}$ with N independent Poisson-commuting integrals $\phi = \{\phi_1, \ldots, \phi_N\}$. If $N_0 = \phi^{-1}(\{0\}) \subset M$ is compact and connected, then*

(1) N_0 is an embedded n-dimensional torus $\mathbb{T}^N$,
(2) There exists an open neighborhood $\mathcal{U}(N_0) \subset M$ of N_0 which can be coordinatized as follows: if $x = \{x_1, \ldots, x_N\}$ are variables on the torus $\mathbb{T}^N = \mathbb{R}^N/\mathbb{Z}^N$ and $y = (y_1, \ldots, y_N) \in D_1$ where $D_1 \subset \mathbb{R}^N$ is a domain containing the origin there exists a diffeomorphism

$$\psi : \mathbb{T}^N \times D_1 \to \mathcal{U}(N_0).$$

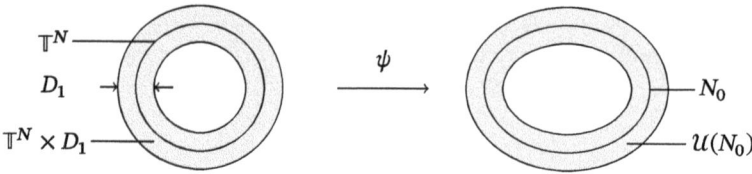

Figure 2.2 Linearizing the flow

Moreover, ψ is symplectic, i.e., $\psi^*\omega = \sum_{i=1}^N dx_i \wedge dy_i$, or equivalently for all smooth functions K, L on M,

$$\{K, L\}_M \circ \psi = \{K \circ \psi, L \circ \psi\}_{(\mathbb{T}^N \times D_1, \sum dx_i \wedge dy_i)},$$

and

$$H \circ \psi = h(y_1, \ldots, y_N) \qquad (2.27)$$

for some function h.

A full proof of the theorem may be found in Moser and Zehnder (2005).

In particular, near a compact, connected level set $\phi^{-1}(0)$, the flow generated by H is extremely simple in terms of the variables $(x, y) \in \mathbb{T}^N \times D_1$ (see below).

Exercise 2.14 Let χ be a *symplectomorphism*, i.e., a symplectic diffeomorphism between symplectic manifolds (M_1, ω_1) and (M_2, ω_2). Let the Hamiltonian $H_2 : M_2 \to \mathbb{R}$, generate the flow $t \mapsto m_2(t)$ on (M_2, ω_2). Show that the Hamiltonian $H_1 = H_2 \circ \chi$ generates the flow $t \mapsto m_1(t) = \chi^{-1}(m_2(t))$ on (M_1, ω_1).

In particular, as $\chi = \psi$ in Theorem 2.13 is a symplectic map from $\mathbb{T}^N \times D_1$ to (M, ω), the flow $m(t)$ generated by H on (M, ω) is transformed to a Hamiltonian flow $(x(t), y(t))$ on $\mathbb{T}^N \times D_1$ generated by the Hamiltonian $h = H \circ \psi$. Thus,

$$\begin{cases} \partial_t x_j = \partial_{y_j}(H \circ \psi) = \partial_{y_j} h, \\ \partial_t y_j = -\partial_{x_j}(H \circ \psi) = -\partial_{x_j} h = 0, \end{cases}$$

so that

$$x_j(t) = x_j(0) + t\partial_{y_j} h(y_1, \ldots, y_N), \qquad y_j(t) = y_j(0). \qquad (2.28)$$

Thus, the system can be integrated explicitly and is given by a straight line motion on a torus. The variables $\{y_j\}$ are called the *actions* of the system, and the $\{x_j\}$ are called the *angles*. Returning to the variables on (M, ω), we have

$$m(t) = \psi\left(x(0) + th_y(y(0)), y(0)\right). \qquad (2.29)$$

The theorem not only tells us how to integrate the system (in terms of the variables x, y which may or may not be hard to construct explicitly) but perhaps more importantly we can understand the qualitative behavior of the system, as we now explain. The quantities

$$w = (w_1, \ldots, w_N) = (\partial_{y_1} h, \ldots, \partial_{y_N} h)$$

are called the *frequencies* of the system. The neighborhood $\mathcal{U}(N_0)$ in the theorem is foliated by tori which we parametrize by the values of y in D_1. As $w = w(y)$, and as y is conserved by the flow, the frequencies are constant on each torus and in general they vary from torus to torus. We distinguish tori according to whether the w_i are rationally independent, i.e., linearly independent over $\mathbb{Q}$, or rationally dependent. In the former case, by a well-known theorem of Kronecker, $\{x_0 + tw : t \in \mathbb{R}\}$ is dense in $\mathbb{T}^N$. Thus, the orbit of the flow is dense on such tori, and in fact the flow is quasi-periodic in time with n frequencies. In the latter case (rational dependence), the flow is restricted to a sub-torus of $\mathbb{T}^N$. For example, if $n = 2$ and $\omega_1 - 2\omega_2 = 0$, then the flow is restricted to

$$\{(x_1, x_2) \in \mathbb{T}^2 : x_1 - 2x_2 = \text{const.} \mod \mathbb{Z}\}.$$

Again, the flow is almost periodic, but with fewer frequencies. Thus, the essential problem of describing the long-time behavior of integrable Hamiltonian systems is solved in principle, provided the invariant set $\phi^{-1}(0)$ is compact and connected. If $\phi^{-1}(0)$ is compact but not connected, we can just restrict our attention to each connected component. Also, if $\phi^{-1}(0)$ is not compact, then the theorem goes through, provided it is known a priori that each ϕ_i generates a global flow, at least for data near $\phi^{-1}(0)$. In this case, one learns that $\phi^{-1}(0)$ has a neighborhood $\mathcal{U}(N_0)$ which is a thickening by an n-dimensional disk D_1 of a product of lines and circles.

$$\mathcal{U}(N_0) = \mathbb{T}^k \times \mathbb{R}^{n-k} \times D_1.$$

On each leaf $\mathbb{T}^k \times \mathbb{R}^{n-k} \times \{y\}$, the flow is again given by a straight line motion, but now the winding takes place on a cylinder rather than a torus.

The LAJ theorem describes the behavior of an integrable system qualitatively. In each case, the difficult task remains to determine the action and angle variables explicitly.

For the convenience of the reader we now give a sketch of the proof of the LAJ theorem.

Sketch of proof. As $\{\phi_i, \phi_j\} = 0$,
$$[v_{\phi_i}, v_{\phi_j}] = v_{\{\phi_j, \phi_i\}} = 0$$
by (2.7), which shows that the Hamiltonian vector fields v_{ϕ_i} induced by the ϕ_i commute, and hence the flows
$$t \to \psi_i^t = \psi_i(t, m), \quad \psi_i(0, m) = m,$$
induced by the v_{ϕ_i} commute (see (2.8)). This means that we can immerse $\mathbb{R}^{2n}$ into $M^{(2n)}$ as follows. Fix $m_0 \in \phi^{-1}(0)$. Then the map
$$t = (t_1, \ldots, t_N) \mapsto T(t) = \psi_1^{t_1} \circ \cdots \circ \psi_N^{t_N}(m_0)$$
takes $\mathbb{R}^N$ into the level set $N_0 = \{m : \phi_i(m) = \phi_i(m_0) \ \forall i = 1, \ldots, n\}$, since
$$\partial_{t_j} \phi_i(\psi_j^{t_j}(m)) = \{\phi_i, \phi_j\}(\psi_j^{t_j}(m)) = 0,$$
so that ϕ_i is a constant of the motion for all the ϕ_j flows, $1 \le i, j \le n$. A simple argument shows that T is onto N_0. Let $\Lambda = \{t \in \mathbb{R}^N : T(t) = m_0\}$. Now, as the flows commute, it follows that Λ is a lattice in $\mathbb{R}^N$. On the other hand, $\mathbb{R}^N/\Lambda$ is mapped diffeomorphically onto N_0, which is compact by assumption. Now the lattice Λ has k generators, $1 \le k \le n$, i.e.,
$$\Lambda = \{j_1 \hat{v}_1 + \cdots + j_k \hat{v}_k : \hat{v}_i \in \mathbb{R}^N\}.$$
If $k < n$, then clearly $N_0 \simeq \mathbb{R}^N/\Lambda$ cannot be compact, which is a contradiction. Hence, we must have $k = n$. But then, $\mathbb{R}^N/\Lambda$ is clearly a torus. Then we must thicken things around N_0, $N_0 \to \mathcal{U}(N_0)$, and so on. □

2.3.1 Remarks on Integrability

As indicated earlier, integrability for dynamical systems has differing connotations in the literature (see, e.g., Beals and Sattinger, 1991). Perhaps the most common notion is LAJ integrability. If the conditions of the LAJ theorem are satisfied, then the flow generated in M^{2N} by the Hamiltonian is equivalent to a straight line motion on a torus $\mathbb{T}^N$, or more generally, on a product of lines and circles. It is in general a very difficult matter to determine the flow explicitly in terms of the original variables in M^{2N}. What one does learn, however, in the toroidal case in particular, is that the flow is almost periodic in time.

Some authors call a dynamical system integrable if it can be written in Lax pair form $\partial_t S = [S, U]$ for some matrices (or more generally, operators, as in the case of the Korteweg–de Vries equation) S and U. As already noted in Chapter 1, Lax pair form implies that the spectrum of S provides S with constants of the motion. If, in particular, S is an $N \times N$ matrix, and the underlying dynamical system has $2N$ degrees of freedom, then the eigenvalues of S, if they are functionally independent and Poisson commute, can be used as ingredients in the LAJ theorem to integrate the system. In the operator theoretic case, such as KdV, the road from Lax pair to integrability is functional analytic and far more complicated (see Gardner et al., 1967).

The mere existence of a Lax pair representation for a dynamical system is, however, no indication at all that the system can be integrated. All it means, in the $N \times N$ matrix case in particular, is that S has N integrals of the motion. And indeed, in a notable recent observation, R. Killip (R. Killip, private communication) showed that any Hamiltonian flow in $\mathbb{R}^{2n}$ which has an equilibrium point, has a Lax pair representation on $n \times n$ matrices.

Exercise 2.15 Provide the details of Killip's construction. Suppose that $H : \mathbb{R}^{2n} \to \mathbb{R}$ is a Hamiltonian with an equilibrium point, located without loss of generality at zero, $\nabla H(0) = 0$. Then the flow generated by H has the form
$$\partial_t z = J \nabla H(z) = JB(z)z,$$
where J is skew symmetric and $B(z)$ is symmetric,
$$J = \begin{pmatrix} 0 & I \\ -I & 0 \end{pmatrix}, \quad B(z) = \int_0^1 H''(\theta z)\, d\theta.$$
Show that the (rank one) matrix $L = zz^T J$ satisfies
$$\partial_t L = [L, -JB].$$
Thus the eigenvalues of L are constants of the motion for this Hamiltonian H. But all but one eigenvalue of L is zero, and this information is not useful.

Actually, from the exercise below, we see that *any* flow in $\mathbb{R}^n$ embeds in a Lax pair.

Exercise 2.16 Extend the construction below, embedding the general flow in $\mathbb{R}^3$, $v' = (x, y, z)' = (f(v), g(v), h(v))$, into 4×4 matrices:

$$\partial_t \begin{pmatrix} 1 & x & 0 & 0 \\ 0 & 2 & y & 0 \\ 0 & 0 & 3 & z \\ 0 & 0 & 0 & 4 \end{pmatrix} = \left[\begin{pmatrix} 1 & x & 0 & 0 \\ 0 & 2 & y & 0 \\ 0 & 0 & 3 & z \\ 0 & 0 & 0 & 4 \end{pmatrix}, \begin{pmatrix} 0 & a & b & d \\ 0 & 0 & c & e \\ 0 & 0 & 0 & k \\ 0 & 0 & 0 & 0 \end{pmatrix} \right]$$

$$= \begin{pmatrix} 0 & -a & -2b-ay+cx & -3d+ex-bz \\ 0 & 0 & -c & -2e+ky-cz \\ 0 & 0 & 0 & -k \\ 0 & 0 & 0 & 0 \end{pmatrix}, \qquad (2.30)$$

where $a = -f$, $c = -g$, $k = -h$ and b, d and e are such that entries $(1, 3), (2, 4)$ and $(1, 4)$ of (2.30) are zero. Show that any flow in $\mathbb{R}^N$ for $N \leq n(n-1)/2$ embeds into $n \times n$ matrices. (Hint for $\mathbb{R}^n$: Note that if L is a $(k+1) \times (k+1)$ upper triangular matrix with distinct nonzero entries on the diagonal, then the map $B \to [L, B]$ from the strictly upper triangular $(k+1) \times (k+1)$ matrices to the strictly upper triangular $(k+1) \times (k+1)$ matrices is a bijection.)

We note here that the fundamental discovery Gardner et al. (1967) that the evolution of initial data of the Korteweg–de Vries equation leads to Schrödinger potentials with the same spectrum (and trivially related reflection coefficients), led to the casting of KdV into Lax pair form (Lax, 1975). It was this new algebraic setup that inspired Flaschka (1974a,b) and Manakov (1974) in their representation of the standard Toda flow (1.19) as an evolution on Jacobi matrices. In 1979, Adler made the crucial observation (discovered independently by Kostant) that the symplectic structure in Chapter 1 for the Toda system, in the center of mass frame, maps symplectically under Flaschka's map $(x_i, y_j) \to (a_i, b_j)$ into the orbit symplectic structure of Kostant–Kirillov. Furthermore, the full Lax pair Toda flow (1.1), (1.2) on general not necessarily tridiagonal matrices, is Hamiltonian with respect to the (natural) Kostant–Kirillov symplectic structure for general orbits (see Section 3.4 for more details). Adler also made the incisive observation that as a consequence of a theorem of Kostant and Symes, the integrability of the (tridiagonal) Toda system follows directly from the splitting of a Lie algebra into a vector space direct sum of Lie algebras. This combination of dynamics and Lie theory is now known collectively as the Adler–Kostant–Symes theorem.

The existence of a Lax pair form is just an invitation to look further for integrability. For example, the Euler–Arnold equation for the rotation of a

2.3 Integration of Vector Fields

rigid body in N dimensions has Lax pair form, but the associated Lax operator S does not have enough independent eigenvalues to integrate the system. However, Manakov (1976) showed how to extend the Lax pair to a new form with enough eigenvalues to integrate the system. As discussed above, the Toda flow on full $N \times N$ symmetric matrices has the Lax pair form (1.1)(1.2), and it follows that the eigenvalues of the Lax operator X provide N integrals for the flow, which is enough to integrate the flow in the tridiagonal case. But, as noted in Section 3.8, for full $N \times N$ matrices, the Toda flow (1.1) takes place on a symplectic manifold of (generic) dimension $2\lfloor N^2/4 \rfloor \gg N$ and so many more integrals of the motion are needed to conclude integrability: This is achieved in Deift et al. (1986) by ad hoc means.

In this monograph, we take a more stringent point of view: By integrability, we mean that the flow can be solved explicitly in the original variables of the system in terms of a well-known, well-understood function theory. An outstanding example of this point of view is the beautiful and deep analysis of KdV in the periodic case (Kappeler and Pöschel, 2003), in which Kappeler and Pöschel construct the Liouville–Arnold tori in an infinite-dimensional phase space. In the case of tridiagonal Toda, the relevant function theory consists of rational functions of exponentials. See, for example, (3.61) and the subsequent equations. Most importantly, these explicit expressions yield the asymptotic behavior of the solutions of the flows as $t \to \infty$ (see (3.78), (3.79)).

Solving the Toda flow asymptotically, however, does not rely on the existence of conserved quantities. This can be done, for example, using Moser's method as in Section 3.2 or using Symes QR factorization method as in Section 3.6. These methods provide information which is pretty much what is needed for algorithmic considerations but do not provide the detailed information in (3.78)–(3.82).

Finally, the Toda flow admits structure which is reminiscent of the one-dimensional heat equation in an interval: In appropriate variables, the vector field decouples into a disconnected set of linear first order differential equations. This property, which we consider in Chapter 4, underlies the so-called *super-integrability* of the system (cf. the comments at the beginning of Section 2.3). Different invariant subsets – the profiles – give rise to a collection of phase spaces in which super-integrability is achieved. For Jacobi matrices, this result was proved in Agrotis et al. (2006). As opposed to action-angle variables, the limit values of orbits also belong to the domain of these new variables, so that asymptotic issues become a matter of local theory. In particular, the precise description of some numerical algorithms concerning QR steps with shifts were obtained in Leite et al. (2010) and Leite et al. (2013).

2.4 Two Classical Examples

We conclude this chapter with two elementary examples of integrable systems: the harmonic oscillator and the simple pendulum (cf. Theorem 2.13 and Figure 2.2).

(1) Harmonic oscillator

$M = \mathbb{R}^2, H = \frac{1}{2}(p^2 + \omega^2 q^2) = \phi_1$ where $\omega \neq 0$. Claim: H is integrable on the (invariant) domain $D = M - \{0\} = \mathbb{R}^2 - \{0\}$. The system is integrable as ϕ_1 is conserved, $n = 1$ and

$$dH = pdp + \omega^2 qdq \neq 0$$

except for $(p, q) = (0, 0)$. Clearly, all nondegenerate Hamiltonians on two-dimensional manifolds are integrable. We have the invariant set

$$N_0 = \{(q, p) : H = \phi_1 = C > 0\} = \{(q, p) : p^2 + \omega^2 q^2 = 2C\},$$

which is clearly a torus. The equations of motion are

$$\partial_t q = \partial_p H = p, \quad \partial_t p = -\partial_q H = -\omega^2 q$$

with solution

$$q = \frac{\sqrt{2C}}{\omega} \sin(\omega t + \alpha), \quad p = \sqrt{2C} \cos(\omega t + \alpha). \tag{2.31}$$

Here α parameterizes the initial data. Note that $p^2 + \omega^2 q^2 = 2C$ and that the set $D = \mathbb{R}^2 \setminus \{0\}$ is invariant under the flow generated by H.

The map ψ in Theorem 2.13 is constructed as follows. Take $D_1 = \mathbb{R}_+$. Then,

$$(y, x) \in \mathbb{R}_+ \times \mathbb{T}^1 \mapsto \psi(y, x) = (q(y, x), p(y, x))$$
$$= \left(\sqrt{\frac{y}{\pi\omega}} \sin(2\pi x), \sqrt{\frac{\omega y}{\pi}} \cos(2\pi x)\right),$$

and $\psi^{-1} : \mathbb{R}^2 \to \mathbb{R}_+ \times \mathbb{T}^1$ takes

$$(q, p) \mapsto (y, x) = \left(\frac{\pi}{\omega}(\omega^2 q^2 + p^2), \frac{1}{2\pi} \sin^{-1}\left(q\sqrt{\frac{\pi\omega}{y}}\right)\right).$$

We have

$$H \circ \psi(y, x) = \frac{1}{2}\left(\left(\sqrt{\frac{\omega y}{\pi}}\cos(2\pi x)\right)^2 + \omega^2\left(\sqrt{\frac{y}{\pi\omega}}\sin(2\pi x)\right)^2\right) = \frac{\omega}{2\pi}y,$$

and

$$\psi^*(dq \wedge dp) = \left(\frac{1}{2\sqrt{\pi\omega y}}\sin(2\pi x)dy + \sqrt{\frac{y}{\pi\omega}}\cos(2\pi x)2\pi dx\right)$$

$$\wedge \left(\frac{1}{2}\sqrt{\frac{\omega}{\pi y}}\cos(2\pi x)dy - \sqrt{\frac{\omega y}{\pi}}\sin(2\pi x)2\pi dx\right)$$

$$= dx \wedge dy.$$

In the (x, y) variables the flow becomes

$$\partial_t x = \partial_y H \circ \psi = \frac{\omega}{2\pi}, \quad \partial_t y = -\partial_x H \circ \psi = 0,$$

so that $x(t) = \frac{\omega t}{2\pi} + x_0$, $y(t) = y_0$, which implies

$$q(t) = \sqrt{\frac{y_0}{\pi\omega}}\sin(\omega t + 2\pi x_0), \quad p(t) = \sqrt{\frac{\omega y_0}{\pi}}\cos(\omega t + 2\pi x_0),$$

as it should (cf. (2.31)).

Exercise 2.17 Verify the formula for ψ^{-1} and show directly that it is symplectic.

(2) Simple pendulum

Here $M^{(2)} = (\mathbb{T} \times \mathbb{R}, \omega = dq \wedge dp)$ and

$$H = \frac{1}{2}p^2 + 1 - \cos(2\pi q),$$

which gives rise to the motion

$$\partial_t q = H_p = p, \quad \partial_t p = -H_q = -2\pi \sin(2\pi q),$$

or

$$\ddot{q} + 2\pi \sin(2\pi q) = 0.$$

Note that for small q, $H \sim \frac{1}{2}p^2 + \frac{1}{2}(2\pi)^2 q^2$, so that the pendulum behaves locally like a simple harmonic oscillator.

Exercise 2.18 The motion of the pendulum described above depends on the value of $H = c > 0$.

1. Show that there are three different cases,

$$c < 2, \quad c = 2, \quad c > 2.$$

- If $c < 2$, the pendulum oscillates with $|2\pi q(t)| \leq \pi$.
- If $c = 2$, the pendulum moves from $2\pi q = -\pi$ to $2\pi q = \pi$ as t runs from $-\infty$ to $+\infty$. This motion is the so-called *separatrix* of the system.
- If $c > 2$, the pendulum rotates "over the top."

2. Describe $N_0 = \{(x, y) : H(x, y) = c\}$ in the above three cases, and draw a picture of $M^{(2)}$ foliated by the invariant sets $N_0 = N_0(c)$ for all values of $c > 0$.

3. Construct the maps ψ and ψ^{-1} of Theorem 2.13 in all three cases.

3
The Toda Lattice

3.1 The Tridiagonal Case

As noted in Chapter 1, the (open) Toda lattice was introduced by H. Toda in 1967 and describes the motion of N particles $(x_i)_{i=1}^N$ on the real line, generated by the Hamiltonian

$$H_{\text{Toda}}(x,y) = \frac{1}{2}\sum_{i=1}^{N} y_i^2 + \sum_{i=1}^{N-1} e^{x_i - x_{i+1}} \tag{3.1}$$

on the symplectic manifold

$$M^{(2N)} = \left(\mathbb{R}^{2N}, \omega = \sum_{i=1}^{N} dx_i \wedge dy_i\right).$$

In other words, the Toda equations are

$$\begin{cases} \partial_t x_i &= \partial_{y_i} H_{\text{Toda}} = y_i, & 1 \leq i \leq N, \\ \partial_t y_i &= -\partial_{x_i} H_{\text{Toda}} = -e^{x_i - x_{i+1}} + e^{x_{i-1} - x_i}, & 2 \leq i \leq N-1, \\ \partial_t y_1 &= -e^{x_1 - x_2}, \\ \partial_t y_N &= e^{x_{N-1} - x_N}. \end{cases} \tag{3.2}$$

As in Flaschka (1974a,b) and Manakov (1974) set

$$\begin{aligned} a_i &:= -y_i/2, & 1 \leq i \leq N, \\ b_i &:= \frac{1}{2} e^{(x_i - x_{i+1})/2}, & 1 \leq i \leq N-1, \end{aligned} \tag{3.3}$$

and consider the real symmetric tridiagonal matrix

$$X = \begin{pmatrix} a_1 & b_1 & & & & \\ b_1 & a_2 & b_2 & & & \\ & b_2 & a_3 & & & \\ & & & \ddots & & \\ & & & & a_{N-1} & b_{N-1} \\ & & & & b_{N-1} & a_N \end{pmatrix} \qquad (3.4)$$

together with the skew symmetric matrix $B(X) = X_- - X_-^T$, where X_- is the strictly lower triangular part of X:

$$B(X) = -B(X)^T = \begin{pmatrix} 0 & -b_1 & & & & \\ b_1 & 0 & -b_2 & & & \\ & b_2 & 0 & & & \\ & & & \ddots & & \\ & & & & 0 & -b_{N-1} \\ & & & & b_{N-1} & 0 \end{pmatrix}. \qquad (3.5)$$

Theorem 3.1 $(x(t), y(t)) \in \mathbb{R}^{2N}$ solves the Toda system (3.2) if and only if

$$\partial_t X = [X, B(X)], \quad X(t=0) = X_0.$$

Proof For $2 \leq i \leq N-1$,

$$\partial_t a_i = -\frac{1}{2}\partial_t y_i = \frac{1}{2}e^{x_i - x_{i+1}} - \frac{1}{2}e^{x_{i-1} - x_i} = 2(b_i^2 - b_{i-1}^2), \qquad (3.6)$$

and for $i = 1, N$, $\partial_t a_1 = 2b_1^2$, $\partial_t a_N = -2b_{N-1}^2$. For $1 \leq i \leq N-1$,

$$\begin{aligned}\partial_t b_i &= \frac{1}{4}e^{\frac{1}{2}(x_i - x_{i+1})}(\partial_t x_i - \partial_t x_{i+1}) \\ &= \frac{1}{2}b_i(y_i - y_{i+1}) = b_i(a_{i+1} - a_i).\end{aligned} \qquad (3.7)$$

Setting $b_0 = b_N \equiv 0$, the Toda equations have the form

$$\begin{cases} \partial_t a_i = 2(b_i^2 - b_{i-1}^2), & 1 \leq i \leq N, \\ \partial_t b_i = b_i(a_{i+1} - a_i), & 1 \leq i \leq N-1. \end{cases} \qquad (3.8)$$

By a simple computation (performed in more generality and detail in Theorem 3.16), the commutator

$$[X, B(X)] = XB - BX = XB + (XB)^T \qquad (3.9)$$

is tridiagonal and its nonzero entries are equal to the evolutions (3.8) of a_i and b_i. In this way we obtain the Lax pair formulation of the Toda system. $\square$

3.1 The Tridiagonal Case

Remark 3.2 Note that the elements on the diagonals $|i-j| = k, 2 \leq k < N$ of $[X, B(X)]$ are automatically zero, so that $[X, B] = [X, B]^T$ is tridiagonal if X is tridiagonal. In the above analysis, this appears to be just a matter of calculation, but actually it has a structural reason: indeed, by construction,

$$B(X) = X + U,$$

where U is upper-triangular. Hence,

$$[X, B(X)] = [X, X] + [X, U] = (XU - UX).$$

It is easy to see that XU and UX have only one nonzero diagonal under the main diagonal, and so as $[X, B(X)]$ is symmetric, it is tridiagonal. Notice the affinity of this argument with the proof in Theorem 1.1 that $X(t)$ is the conjugation of $X(0)$ by orthogonal matrices (and hence preserves symmetry) and by upper triangular matrices (and hence preserves the zero entries below the nonzero diagonal under the main diagonal).

Exercise 3.3 Use the above argument to show that if X is a symmetric banded matrix, then $[X, B(X)]$ has the same band structure as X.

Thus, the Toda flow is band-preserving for general symmetric banded matrices, not just tridiagonal matrices. We will say more about this later (see Remark 3.24 and also Chapter 4).

Now note that

$$H_{\text{Toda}}(x, y) = 2 \sum_{i=1}^{N} a_i^2 + 4 \sum_{i=1}^{N-1} b_i^2,$$

i.e.,

$$H_{\text{Toda}}(x, y) = 2 \operatorname{tr}(X^2). \tag{3.10}$$

As $\operatorname{tr} X^2$ is conserved, we have, as noted in Chapter 1, an a priori bound on solutions of (3.8), $\operatorname{tr}(X^2) = \operatorname{tr}(X_0^2)$, so that by general ODE theory the solutions of the Toda equations have global existence. Note that, as

$$b_i^2(t) = \frac{1}{4} e^{(x_i - x_{i+1})} \leq \operatorname{tr}(X_0^2),$$

we see a priori that

$$x_i(t) \leq x_{i+1}(t) + c \tag{3.11}$$

for some $c < \infty$, so that the particle i cannot get too far ahead of the particle $i + 1$. We will see, in fact, that the particles will order themselves as $x_1(t) < x_2(t) < \cdots < x_N(t)$, as $t \to \infty$. Note also that a direct proof of global existence for the Toda equations in (x, y) variables proceeds by noting that

$$\frac{1}{2}\sum_{i=1}^{N} y_i^2 \leq H_{\text{Toda}}(x, y) = \text{const.},$$

so that

$$|y_i(t)| \leq 2H_{\text{Toda}}^{1/2},$$

and consequently, writing $x_i(t) = x_i(0) + \int_0^t y_i(s)ds$,

$$|x_i(t)| \leq |x_i(0)| + 2tH_{\text{Toda}}^{1/2},$$

from which standard ODE arguments imply global existence, even though the x_i grow (linearly) as $t \to \infty$.

As noted in Chapter 1, the Lax pair form of the Toda equations implies that

$$\text{spec}(X(t)) = \text{spec}(X_0).$$

Thus

$$\lambda_k(t) = \lambda_k(0), \qquad \forall k = 1, \ldots, N, \tag{3.12}$$

where the λ_k are the eigenvalues of $X(t)$. We will show later on that they are independent and Poisson-commute. Our immediate goal is to solve the Toda equations explicitly.

We need some preliminary results which are also of general interest.

Definition 3.4 A real, symmetric, tridiagonal matrix with strictly positive off-diagonal entries is called a *Jacobi matrix*.

Notice that, from equation (3.3), the b_i are exponentials, and thus are positive. The matrices $X(t)$ above are thus Jacobi matrices, and we will denote them by J (not to be confused with the use of J in Section 2.1) to distinguish the tridiagonal and the full matrix cases. Recall that for real symmetric matrices, an eigenvalue is algebraically simple if and only if it is geometrically simple.

Lemma 3.5 *Let J be a Jacobi matrix. Then the spectrum of J is simple, and the first and last components of eigenvectors of J are nonzero, i.e., if $Jv = \lambda v$, $v = (v_1, \ldots, v_N)^T \neq 0$, then $v_1, v_N \neq 0$.*

3.1 The Tridiagonal Case

Proof Let u be an eigenvector associated to $\lambda \in \text{spec}(J)$. Then

$$\begin{pmatrix} a_1 - \lambda & b_1 & & & \\ b_1 & a_2 - \lambda & \ddots & & \\ & \ddots & \ddots & b_{N-1} & \\ & & b_{N-1} & a_N - \lambda \end{pmatrix} \begin{pmatrix} v_1 \\ v_2 \\ \vdots \\ v_N \end{pmatrix} = 0,$$

which implies

$$(a_1 - \lambda)v_1 + b_1 v_2 = 0,$$
$$b_1 v_1 + (a_2 - \lambda)v_2 + b_2 v_3 = 0,$$
$$\vdots$$
$$b_{k-1} v_{k-1} + (a_k - \lambda)v_k + b_k v_{k+1} = 0,$$
$$\vdots$$
$$b_{N-1} v_{N-1} + (a_N - \lambda)v_N = 0.$$

Thus if $v_1 = 0$, we see from the first equation that, as $b_1 \neq 0$, $v_2 = 0$, and so on. A straightforward recurrence leads to $v = 0$, which is a contradiction. So $v_1 \neq 0$; the same is true for v_N by a similar argument. Now, if λ is not a simple eigenvalue, then there exist at least two eigenvectors v, w that are linearly independent. Every combination $v + \alpha w$ is also an eigenvector with eigenvalue λ; choosing α so that $(v + \alpha w)_1 = 0$ leads to $v + \alpha w = 0$, a contradiction. It follows that the spectrum of J is simple. □

The following result is basic in perturbation theory (see, e.g., Reed and Simon, 1978 or Kato, 1995).

Theorem 3.6 *Suppose A_0 is an $N \times N$ (complex) matrix and that λ_0 is an algebraically, and hence geometrically, simple eigenvalue of A_0, $A_0 v_0 = \lambda_0 v_0$, $v_0 \neq 0$. Then, for A in a neighborhood $S_\delta = \{A \mid \|A - A_0\| < \delta\}$ of A_0 for some $\delta > 0$, A has an algebraically simple eigenvalue $\lambda(A)$ which is analytic in the entries of $\{a_{ij}\}$ of A in S_δ, where $\lambda(A) \to \lambda_0$ as $\|A - A_0\| \to 0$. Furthermore, the eigenvector associated with $\lambda(A)$, $Av(A) = \lambda(A)v(A)$, may be chosen as an analytic function of the entries of $A \in S_\delta$, and $v(A) \to v_0$ as $\|A - A_0\| \to 0$. Finally, if $A_0, A \in S_\delta$ are real symmetric, then $v(A)$ is real analytic and if v_0 is normalized, then $v(A)$ may be chosen to be normalized, $\|v(A)\| = \|v_0\| = 1$.*

Proof We provide a sketch of the proof. Choose $\epsilon > 0$ sufficiently small so that the only eigenvalue of A_0 in $\{z \mid |z - \lambda_0| \leq \epsilon\}$ is λ_0. Then, for $\delta > 0$ sufficiently small, set

$$P(A) = \frac{1}{2\pi i} \int_{C_\epsilon} \frac{dz}{z - A}, \quad A \in S_\delta,$$

where C_ϵ is the positively oriented circle $|z - \lambda_0| = \epsilon$. Then $P(A)$ is a rank one projection which commutes with A, and is analytic in the entries of A. Set $v(A) = P(A)v_0$. Then

$$Av(A) = AP(A)v_0 = P(A)Av_0 = \lambda(A)v(A)$$

for some scalar $\lambda(A)$, as $P(A)$ is rank one. It follows that

$$\lambda(A) = \langle v_0, Av(A)\rangle / \langle v_0, v(A)\rangle.$$

Thus A has an analytic eigenvalue $\lambda(A)$ and associated eigenvector $v(A)$, for which $\lambda(A_0) = \lambda_0$ and $v(A_0) = v_0$.

If A_0 and A are real symmetric, then $P(A)$ is real symmetric and

$$\|v(A)\|^2 = \langle P(A)v_0, P(A)v_0\rangle = \langle v_0, P^2(A)v_0\rangle = \langle v_0, P(A)v_0\rangle,$$

which implies that $\|v(A)\| = \sqrt{\langle v_0, P(A)v_0\rangle}$ is real analytic. Hence $\hat{v}(A) = v(A)/\sqrt{\langle v_0, P(A)v_0\rangle}$ is normalized and real analytic. □

Exercise 3.7 Show that if A is a real, symmetric matrix with a simple eigenvalue $\lambda(A)$ and normalized eigenvector $\hat{v}(A)$ as in the proof of Theorem 3.6, then if $A(t)$, $A(t = 0) = A_0$, depends smoothly on some real parameter t, $|t| < \epsilon$ for some small $\epsilon > 0$, then

$$\frac{d}{dt}\lambda(A(t)) = \left(\hat{v}\left(A(t), \left(\frac{d}{dt}A(t)\right)\hat{v}(A(t))\right)\right).$$

In particular, show that if A is a Jacobi matrix, as in (3.4), then

$$\frac{\partial}{\partial a_i}\lambda(A) = \hat{v}_i^2(A), \quad 1 \le i \le N,$$

and

$$\frac{\partial}{\partial b_j}\lambda(A) = 2\hat{v}_j(A)\hat{v}_{j+1}(A), \quad 1 \le j \le N-1,$$

where $\hat{v}(A) = (\hat{v}_1(A), \dots, \hat{v}_N(A))^T$.

Let $\mathcal{J} = \mathcal{J}_N$ denote the collection of $N \times N$ Jacobi matrices. For a matrix $J \in \mathcal{J}$, we may order the eigenvalues

$$\lambda_1 > \cdots > \lambda_N$$

and specify normalized eigenvectors $Ju_i = \lambda_i u_i$, $u_i = (u_{i,1}, \dots, u_{i,N})^T$, $\|u_i\|^2 = 1$, uniquely, by requiring that their first components be positive, i.e., $u_{i,1} > 0, 1 \le i \le N$. With these specifications, we have the following basic result.

3.1 The Tridiagonal Case

Proposition 3.8 *Let $N > 1$ and*

$$M = \{(\alpha, \beta) = (\alpha_1, \ldots, \alpha_N, (\beta_1, \ldots, \beta_N)^T) :$$
$$\alpha_1 > \cdots > \alpha_N, \sum \beta_i^2 = 1, \beta_i > 0\}.$$

The map $\psi : \mathcal{J} \to M$, $J \mapsto (\lambda, u) = (\lambda_1, \ldots, \lambda_N, (u_{1,1}, \ldots, u_{N,1})^T)$ is a diffeomorphism.

Part of the proof follows an argument in Gragg and Harrod (1984).

Proof We first show surjectivity: given $(\lambda, u) \in M$, we obtain $J \in \mathcal{J}$ such that $\psi(J) = (\lambda, u)$. For such a matrix J, the spectral theorem would imply

$$J = Q^T \Lambda Q, \quad (3.13)$$

where $\Lambda = \text{diag}(\lambda_1, \ldots, \lambda_N)$, $\lambda_1 > \cdots > \lambda_N$, and an orthogonal matrix Q with first column equal to u. We show how to compute Q.

Let K consist of the matrix with columns $u, \Lambda u, \ldots, \Lambda^{N-1} u$ and V be the Vandermonde matrix with columns $\mathbf{1}, \Lambda \mathbf{1}, \ldots, \Lambda^{N-1} \mathbf{1}$, were $\mathbf{1} \in \mathbb{R}^N$ is the column vector with entries equal to 1. Clearly $\det K = \Pi_{i=1}^N u_{i,1} \det V$: by the properties of λ and u, $\det K$ is necessarily nonzero. In particular, there is a unique QR decomposition $K = \hat{Q} R$, where the diagonal entries of R are strictly positive. Notice that the first columns of K and $\hat{Q}$ are equal to u.

We first show that $T := \hat{Q}^T \Lambda \hat{Q}$ is a symmetric tridiagonal matrix. Symmetry is clear: Tridiagonality follows if we prove that T is also an upper Hessenberg matrix, i.e., a matrix H for which $H_{i,j} = 0$, $i = 3, \ldots, N$, $j = 1, \ldots i - 2$.

Applying Λ to the ith column of K gives its $(i+1)$th columns. In matrix notation, $\Lambda K = KC$, where C is a companion matrix: The first $N-1$ columns of C are the canonical vectors $e_2, \ldots, e_N$, and the last column is a linear combination of the first $N - 1$ columns, by the Cayley–Hamilton theorem. As $K = \hat{Q} R$, we have $T = RK^{-1} \Lambda K R^{-1} = RCR^{-1}$, necessarily an upper Hessenberg matrix.

The matrix T however is not necessarily a Jacobi matrix. The off-diagonal entries $T_{i,i-1}$ are nonzero, as $T = RCR^{-1}$. It is just a matter of adjusting their signs. We look for a diagonal matrix $E = E^{-1}$ with $E_{1,1} > 0$ such that $J = ETE$ is a Jacobi matrix — this is done sequentially (the sign of E_{22} is determined by the known sign of E_{11} and the requirement that $(ETE)_{21} > 0$).

We now consider the injectivity of ψ. For $J \in \mathcal{J}$, let $\hat{J}$ be the matrix be obtained by the construction above from datum $(\lambda, u) = \psi(J)$: we show that $J = \hat{J}$. From the spectral theorem, $J = Q^T \Lambda Q$ and $\hat{J} = \hat{Q}^T \Lambda \hat{Q}$, where the first columns of Q and $\hat{Q}$ equal the same vector u^T with positive entries. Then $(\hat{Q} Q^T) \Lambda = \Lambda (\hat{Q} Q^T)$ and, since Λ has simple spectrum, $(\hat{Q} Q^T)$ must be a function of Λ. Thus $(\hat{Q}Q^T)$ is an orthogonal diagonal matrix F. In matrix notation, $\hat{Q} = FQ$, and since the first columns of $\hat{Q}$ and Q are the same, $F = I$ and $J = \hat{J}$.

Clearly ψ is smooth by Theorem 3.6. Its inverse ϕ is also smooth as $(\lambda, u) \to K \to \hat{Q} \to T \to \hat{J}$ is smooth. As $\psi \circ \phi = \text{id}$, its Jacobians are invertible: thus both maps are diffeomorphisms. □

Let e_i be the canonical vectors of $\mathbb{R}^N$. We now show how the first components

$$(u_{11}(t), \ldots, u_{1N}(t))$$

of the normalized eigenvectors of $X(t)$ evolve under the Toda flow. As the eigenvalues $\lambda_k(t), 1 \le k \le N$ are constant under the flow, the evolution of the $u_{1j}(t)$ completely determines the evolution of $J(t)$ through the inversion

$$J(t) = \psi^{-1}(\lambda_1, \ldots, \lambda_N, (u_{11}(t), \ldots, u_{1N}(t))^T). \tag{3.14}$$

Theorem 3.9 (Moser) *Let $J = J(t) \in \mathcal{J}$ be a solution of the Toda flow,*

$$\partial_t J = [J, B(J)]$$

with $J(0) = J_0$. Let

$$\psi(J(t)) = (\lambda_1(t), \ldots, \lambda_N(t), u_{11}(t), \ldots, u_{1N}(t))$$

be the eigenvalues and first components of the normalized eigenvectors u_i of $J(t)$.

Then for $t > 0$ and $1 \le i \le N$,

$$\lambda_i(t) = \lambda_i(0), \tag{3.15}$$

and

$$u_{1,i}(t) = \frac{u_{1,i}(0)e^{\lambda_i t}}{\left(\sum_{j=1}^N u_{1j}^2(0)e^{2\lambda_j t}\right)^{1/2}}. \tag{3.16}$$

Proof Equation (3.15) is already proven. As φ is a diffeomorphism and $J(t)$ is a smooth function of t, $u_i(t)$ is a smooth function of t. Differentiating

$$\begin{aligned}(J(t) - \lambda_i(t))u_i(t) &= 0, \\ u_i(t) &= (u_{1,i}(t), \ldots, u_{N,i}(t))^T, \quad u_{1,i}(t) > 0,\end{aligned} \tag{3.17}$$

we obtain

$$(\partial_t J - \partial_t \lambda_i)u_i(t) + (J - \lambda_i)\partial_t u_i = 0,$$

and so

$$(JB(J) - B(J)J)u_i(t) + (J - \lambda_i)\partial_t u_i = 0,$$

3.1 The Tridiagonal Case

leading to

$$(J - \lambda_i)(B(J)u_i(t) + \partial_t u_i) = 0.$$

As the λ_i are simple, we must have

$$B(J)u_i(t) + \partial_t u_i(t) = \alpha(t)u_i(t)$$

for some $\alpha(t)$. Taking inner product with $u_i(t)$, we find

$$\alpha(t) = (u_i(t), B(J)u_i(t)) + (u_i(t), \partial_t u_i(t)).$$

But $\|u_i(t)\|^2 = 1$, which implies $(u_i(t), \partial_t u_i(t)) = 0$, and $(u_i(t), B(J)u_i(t)) = 0$, as $B(J)$ is skew symmetric. Thus, $\alpha(t) = 0$. We conclude that $\partial_t u_i(t) = -B(J)u_i(t)$. In particular,

$$\partial_t u_{1,i}(t) = -(e_1, B(J)u_i(t)) = J_{12} u_{2,i}(t).$$

But as $(J - \lambda_i)u_i = 0$, we have in particular

$$J_{11} u_{1,i} - \lambda_i u_{1,i} + J_{12} u_{2,i} = 0,$$

which implies that

$$\partial_t u_{1,i}(t) = (\lambda_i - J_{11})u_{1,i}(t). \tag{3.18}$$

Diagonalize $J = U\Lambda U^T$. Now,

$$J_{11} = \langle e_1, U\Lambda U^T e_1 \rangle = \sum_{i=1}^{N} \lambda_i u_{1,i}^2(t). \tag{3.19}$$

Thus, we have the system of equations for $u_{11}(t), \ldots, u_{1N}(t)$:

$$\partial_t u_{1,i} = \left(\lambda_i - \sum_{j=1}^{N} \lambda_j u_{1,j}^2(t) \right) u_{1,i}(t). \tag{3.20}$$

It follows that, for $1 \leq i, k \leq N$,

$$\partial_t \ln\left(\frac{u_{1k}(t)}{u_{1,i}(t)} \right) = \lambda_k - \lambda_i, \tag{3.21}$$

so that

$$\frac{u_{1k}(t)}{u_{1,i}(t)} = \frac{u_{1k}(0)}{u_{1,i}(0)} e^{(\lambda_k - \lambda_i)t}. \tag{3.22}$$

Using $\sum_k u_{1k}^2 = 1$, we find from (3.22) that

$$\frac{1}{u_{1,i}^2(t)} = \frac{1}{u_{1,i}^2(0)} \sum_{k=1}^{N} u_{1k}^2(0) e^{2(\lambda_k - \lambda_i)t},$$

or

$$u_{1,i}(t) = \frac{u_{1,i}(0)}{\left(\sum_{k=1}^{N} u_{1k}^2(0)e^{2(\lambda_k - \lambda_i)t}\right)^{1/2}} = \frac{u_{1,i}(0)e^{\lambda_i t}}{\left(\sum_{k=1}^{N} u_{1k}^2(0)e^{2\lambda_k t}\right)^{1/2}}. \quad (3.23)$$

This proves the theorem. □

Remark 3.10 The argument above is from Moser (1975b). A similar computation for the full Toda flow on $N \times N$ symmetric matrices (Theorem 3.37) yields the same formula (3.16), suitably interpreted.

3.2 Long-Time Behavior

We now begin to analyze the long-time behavior of the Toda flow. Order the eigenvalues:

$$\lambda_1 > \cdots > \lambda_N.$$

Then, as $u_{1k}(0) > 0$ for all k, it follows from (3.23) that

$$u_{1i}(t) = \frac{u_{1,i}(0)e^{(\lambda_i - \lambda_1)t}}{\left(u_{11}^2(0) + \sum_{j=2}^{N} u_{1,j}^2(0)e^{2(\lambda_j - \lambda_1)t}\right)^{1/2}}, \quad (3.24)$$

so that, as $t \to \infty$,

$$u_{1i}(t) \to \delta_{1,i} \quad (3.25)$$

exponentially fast. Hence, with $u = (u_{1,1}, \ldots, u_{N,1})^T$, as $t \to \infty$,

$$a_1(t) = \langle u, \Lambda u \rangle = \sum_{i=1}^{N} \lambda_i u_{1,i}^2(t) \to \lambda_1. \quad (3.26)$$

As

$$u_{11}^2(t) = \frac{u_{11}(0)^2}{\left(u_{11}^2(0) + \sum_{j=2}^{N} u_{1j}^2(0)e^{2(\lambda_j - \lambda_1)t}\right)^{1/2}},$$

we see that

$$u_{11}^2(t) = 1 + O(e^{2(\lambda_2 - \lambda_1)t}). \quad (3.27)$$

Inserting this relation into the above formula for $a_1(t)$ and using (3.25) we see that in fact $a_1(t) \to \lambda_1$ exponentially fast,

$$a_1(t) = \lambda_1 + O(e^{2(\lambda_2 - \lambda_1)t}). \quad (3.28)$$

3.2 Long-Time Behavior

Now, as $\langle u, \Lambda^2 u \rangle = (J^2)_{11} = a_1^2 + b_1^2$,

$$b_1^2 = \sum_{i=1}^{N} \lambda_i^2 u_{1,i}^2 - a_1^2 = \sum_{i=2}^{N} (\lambda_i - \lambda_1)^2 u_{1,i}^2 - (a_1 - \lambda_1)^2. \tag{3.29}$$

So from (3.24), (3.25), and (3.28) we have $b_1(t) \to 0$ exponentially fast as $t \to \infty$,

$$b_1^2(t) = (\lambda_2 - \lambda_1)^2 \frac{u_{1,2}^2(0)}{u_{1,1}^2(0)} e^{2(\lambda_2 - \lambda_1)t}(1 + o(1)), \tag{3.30}$$

where the term o(1) is exponentially small as $t \to \infty$. (cf. (3.67)). From (3.28), $J_{11} = a_1$ converges to the top eigenvalue of $J(0)$ exponentially fast as $t \to -\infty$. In terms of the original Toda variables,

$$y_1 = -2a_1.$$

Thus

$$\partial_t x_1 = y_1 = -2a_1 = -2\lambda_1 + O(e^{2(\lambda_2 - \lambda_1)t}),$$

and we conclude that

$$x_1(t) = c_1 - 2\lambda_1 t + O(e^{2(\lambda_2 - \lambda_1)t}) \tag{3.31}$$

for some constant c_1, so that $x_1(t)$ moves asymptotically linearly with velocity $-2\lambda_1$ as $t \to \infty$. On the other hand, $b_1 = \frac{1}{2}e^{\frac{1}{2}(x_1 - x_2)}$ and hence

$$e^{(x_1(t) - x_2(t))} = O(e^{2(\lambda_2 - \lambda_1)t})$$

so that, as $t \to \infty$, $x_1(t)$ lies to the left of $x_2(t)$. More precisely, from (3.30), it follows that, as $t \to \infty$,

$$x_1(t) - x_2(t) = 2(\lambda_2 - \lambda_1)t + O(1), \tag{3.32}$$

and then from (3.31),

$$x_2(t) = -2\lambda_2 t + O(1). \tag{3.33}$$

We note that the gap between $x_2(t)$ and $x_1(t)$ increases linearly.

One can in principle continue in this way to determine the asymptotic behavior of $x_3(t), \ldots, x_N(t)$, but this approach is cumbersome. We now present a different approach, also due to Moser, which shows quite simply – but without a precise rate of convergence – that as $t \to \infty$,

$$a_k(t) \to \lambda_k, \quad b_k(t) \to 0. \tag{3.34}$$

One can then conclude that,

$$x_k = -2t\lambda_k + O(1), \quad 1 \le k \le N. \tag{3.35}$$

Theorem 3.11 (Moser) *Let $J(t)$ solve the Toda equations with $X(0) = X_0$, a $N \times N$ Jacobi matrix. Then*

$$J(t) \xrightarrow[t \to \infty]{} \operatorname{diag}(\lambda_1, \ldots, \lambda_N), \tag{3.36}$$

*where $\lambda_1 > \cdots > \lambda_N$ are the eigenvalues of J_0. In particular, the Toda flow on Jacobi matrices is an **ordering** eigenvalue algorithm.*

Proof From (3.8),

$$\partial_t a_i = 2\left(b_i^2 - b_{i-1}^2\right), \quad \partial_t b_i = b_i(a_{i+1} - a_i),$$

we have $\partial_t a_1 = 2b_1^2 > 0$, so that $a_1(t)$ is increasing. Integrating, we obtain

$$a_1(t) = a_1(0) + 2 \int_0^t b_1^2(s) ds,$$

as

$$\sum_{i=1}^{N} a_i^2 + 2 \sum_{i=1}^{N-1} b_i^2 = \frac{1}{2} H_{\text{Toda}} = \text{const.}, \tag{3.37}$$

we conclude, in particular, that $a_1(t)$ is bounded, and hence

$$a_1(\infty) = \lim_{t \to \infty} a_1(t)$$

exists and is finite. Thus $b_1^2(t)$ is integrable and

$$a_1(\infty) = a_1(0) + 2 \int_0^\infty b_1^2(s) ds. \tag{3.38}$$

As $\partial_t b_1 = b_1(a_2 - a_1)$, and as a_2, a_1 and b_1 are also bounded, it follows that $\partial_t b_1$ is bounded. In particular, $b_1(t)$ is uniformly continuous, hence as $\int_0^\infty b_1^2(s) < \infty$, an elementary calculation implies that

$$b_1(t) \to 0,$$

as $t \to \infty$. Now,

$$\partial_t (a_1 + a_2) = 2b_2^2 > 0, \tag{3.39}$$

and as $a_1 + a_2$ is bounded, monotonicity again implies that $\lim a_1 + a_2$ exists. But we have already showed that $\lim a_1$ exists, and so

$$a_2(\infty) = \lim_{t \to \infty} a_2(t)$$

exists and is finite. But integrating (3.39) as in (3.38), we now conclude that

$$b_2(t) \to 0,$$

as $t \to \infty$. Proceeding by induction using

$$\partial_t(a_1 + \cdots + a_k) = 2b_k^2, \quad 1 \le k \le N-1,$$

we conclude that

$$a_k(\infty) = \lim_{t \to \infty} a_k(t), \quad 1 \le k \le N, \qquad (3.40)$$

exists, and as $t \to \infty$,

$$b_k(t) \to 0, \quad 1 \le k \le N-1. \qquad (3.41)$$

As the Toda flow is isospectral, the $a_k(\infty)$ must be the eigenvalues of J_0. But which eigenvalue corresponds to a given $a_k(\infty)$? Arguing as above, we have

$$\partial_t x_k = y_k = -2a_k = -2a_k(\infty) + o(1)$$

and so, as $t \to \infty$,

$$x_k(t) = -2a_k(\infty)t + o(t), \quad 1 \le k \le N, \qquad (3.42)$$

and so,

$$x_k(t) - x_{k+1}(t) = -2(a_k(\infty) - a_{k+1}(\infty))t + o(t). \qquad (3.42')$$

As the flow is isospectral we have for $\lambda_k \in \sigma(J(0)) = \sigma(J(t)), \det(J(t)-\lambda_k) = 0$), and so as $b_j(t) \to 0$ and $t \to \infty, 1 \le j \le N-1$,

$$\det(a_1(\infty) - \lambda_k, \ldots, a_N(\infty) - \lambda_k) = \lim_{t \to \infty} \det J(t) - \lambda_k) = 0.$$

It follows that $\lambda_k = a_{\pi(k)}(\infty), 1 \le k \le N$ for some permutation $\pi \in S_N$. In particular, as the $\lambda_k's$ are distinct, the $a_j's$ are distinct, and so as $b_j = \frac{1}{2}e^{\frac{1}{2}(x_j - x_{j+1})} \to 0$ as $t \to \infty$, it follows from (3.42)' that

$$a_1(\infty) > a_2(\infty) \ldots > a_N(\infty).$$

We conclude that

$$a_k(\infty) = \lambda_k, \quad 1 \le k \le N. \qquad (3.43)$$

Now, as $\partial_t a_1(t) = 2b_1^2(t)$, we see that $a_1(t)$ is monotone decreasing as $t \to -\infty$, so that, as above,

$$a_1(-\infty) = \lim_{t \to -\infty} a_1(t) \qquad (3.44)$$

exists and $b_1(t) \to 0$. Continuing, we find that

$$a_k(-\infty) = \lim_{t \to -\infty} a_k(t), \quad 1 \le k \le N, \qquad (3.45)$$

exists, and

$$b_k(t) \xrightarrow[t \to -\infty]{} 0. \qquad (3.46)$$

Necessarily, the $a_k(-\infty)$ are the eigenvalues of J_0. Arguing as above, we have

$$x_k(t) = -2a_k(-\infty)t + o(t), \qquad (3.47)$$

and so for $k = 1, \ldots, N-1$,

$$x_k(t) - x_{k+1}(t) = -2(a_k(-\infty) - a_{k+1}(-\infty))t + o(t). \qquad (3.48)$$

But again, as $b_k = \frac{1}{2}e^{\frac{1}{2}(x_k - x_{k+1})} \xrightarrow[t \to -\infty]{} 0$, we must have

$$a_k(-\infty) - a_{k+1}(-\infty) < 0.$$

It follows that we must have

$$a_k(-\infty) = \lambda_{N-k+1}, \quad 1 \le k \le N. \qquad (3.49)$$

$\square$

Thus, we have a "billiard ball"-type interaction: as t increases from $-\infty$,

$$\begin{array}{cccc} -2\lambda_N t & -2\lambda_{N-1} t & -2\lambda_2 t & -2\lambda_1 t \\ * & \longleftarrow\!\!* & \cdots \quad \longleftarrow\!\!* & \longleftarrow\!\!* \\ x_1 & x_2 & x_{N-1} & x_N \end{array},$$

where the arrows indicate the relative speed of the particles in a reference frame with origin at x_1, and as $t \to +\infty$,

$$\begin{array}{cccc} -2\lambda_1 t & -2\lambda_2 t & -2\lambda_{N-1} t & -2\lambda_N t \\ \longleftarrow\!\!* & \longleftarrow\!\!* & \cdots \quad \longleftarrow\!\!* & * \\ x_1 & x_2 & x_{N-1} & x_N \end{array},$$

where the arrows now represent the relative speed with respect to x_N; so that the particle x_N, which has velocity $-2\lambda_1$ as $t \to -\infty$, transfers this velocity to x_1 as $t \to +\infty$, and so on. This is reminiscent of Newton's cradle (see Figure 3.1), where a ball impinging on a row of balls with velocity v transfers its velocity to the end ball after collision.

Finally, observe that from (3.42) and (3.47), $b_k(t) \to 0$ exponentially fast as $t \to \pm\infty$.

3.2 Long-Time Behavior

Figure 3.1 Newton's cradle. In an ideal setting (the absence of friction of any kind, perfect alignment, etc.), the first ball transfers its velocity to the last ball.

We now obtain more detailed information on the error terms $o(t)$ in equations (3.42) and (3.47). In a series of steps, we show that, for $1 \leq k \leq N$,

$$x_k(t) = -2t\lambda_k + \beta_k^+ + o(1) \qquad \text{as } t \to +\infty, \tag{3.50}$$

and

$$x_k(t) = -2t\lambda_{N-k+1} + \beta_k^- + o(1) \qquad \text{as } t \to -\infty, \tag{3.51}$$

where the expressions for β_k^+ and β_k^- are given in equations (3.80) and (3.81).

We will also derive the remarkable formula (3.82) of Moser and Zehnder for the *phase shifts* $\beta_k^+ - \beta_{N-k-1}^-$ relating $J(t = \infty)$ to $J(t = -\infty)$. Consider

$$x_k(t) \sim -2\lambda_k t \text{ as } t \to +\infty,$$
$$x_{N-k+1}(t) \sim -2\lambda_k t \text{ as } t \to -\infty.$$

Thus, the particle x_{N-k+1} traveling with velocity $-2\lambda_k$ as $t \to -\infty$, somehow "transfers" its velocity to x_k after collision, as $t \to +\infty$. This is why the phase shift is defined as $\beta_k^+ - \beta_{N-k+1}^-$, as depicted in Figure 3.2.

Using *Z-variables* in place of the data $\psi(J)$ in Proposition 3.8, we give another derivation of the above phase shifts in Chapter 4 using Theorem 4.18, following Leite et al. (2008).

Step 1: Expressing the entries b_k in terms of $\psi(J)$, equation (3.60).

From the theory of skew-symmetric tensors, for $u_0, u_1, \ldots, u_k \in \mathbb{R}^N$,

$$\langle u_0 \wedge \cdots \wedge u_k, u_0 \wedge \cdots \wedge u_k \rangle = \det\left(\langle u_i, u_j \rangle\right)_{0 \leq i,j \leq k}. \tag{3.52}$$

In particular, for

$$u_i = J_0^i e_1, \quad 0 \leq i \leq k, \tag{3.53}$$

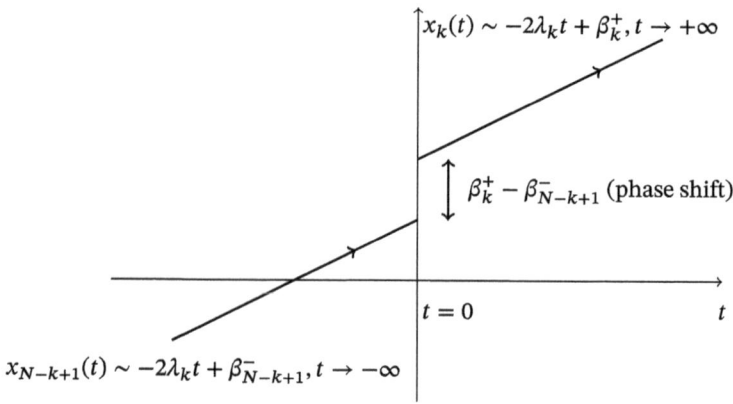

Figure 3.2 The phase shifts.

where $J_0 = J(0)$ and $e_1 = (1, 0, \ldots, 0)^T$,

$$\langle e_1 \wedge J_0 e_1 \wedge \cdots \wedge J_0^k e_1, e_1 \wedge J_0 e_1 \wedge \cdots \wedge J_0^k e_1 \rangle$$

$$= \det \begin{pmatrix} \langle e_1, e_1 \rangle & \langle e_1, J_0 e_1 \rangle & \cdots & \langle e_1, J_0^k e_1 \rangle \\ \langle J_0 e_1, e_1 \rangle & \langle J_0 e_1, J_0 e_1 \rangle & \cdots & \langle J_0 e_1, J_0^k e_1 \rangle \\ \vdots & & & \\ \langle J_0^k e_1, e_1 \rangle & \cdots & & \langle J_0^k e_1, J_0^k e_1 \rangle \end{pmatrix}$$

$$= \det \begin{pmatrix} c_0 & c_1 & \cdots & c_k \\ c_1 & c_2 & \cdots & \\ \vdots & & & \\ c_k & \cdots & & c_{2k} \end{pmatrix} = D_k,$$

where

$$c_j = \langle J_0^j e_1, e_1 \rangle = \sum_{i=1}^N \lambda_i^j u_i(1)^2, \tag{3.54}$$

and $\{\lambda_i\}$ are the eigenvalues of J_0 and $u_i(1) > 0$ are the first components of the corresponding eigenvectors. Rewriting, we have

$$c_j = \int \lambda^j d\mu(\lambda) \tag{3.55}$$

for the *spectral measure*

$$d\mu(\lambda) = \sum_{i=1}^N u_i^2(1) \delta_{\lambda_i}(\lambda). \tag{3.56}$$

3.2 Long-Time Behavior

On the other hand, as

$$J_0 = \begin{pmatrix} a_1 & b_1 & 0 & \cdots & 0 \\ b_1 & a_2 & b_2 & \ddots & \vdots \\ 0 & b_2 & \ddots & \ddots & 0 \\ \vdots & \ddots & \ddots & & b_{N-1} \\ 0 & \cdots & 0 & b_{N-1} & a_N \end{pmatrix}$$

is a Jacobi matrix,

$$J_0 e_1 = b_1 e_2 + a_1 e_1 = b_1 e_2 + r_1, \quad r_1 \in \text{Span}(e_1),$$
$$J_0^2 e_1 = b_1 J_0 e_2 + a_1 J_0 e_1 = b_1 b_2 e_3 + r_2, \quad r_2 \in \text{Span}(e_1, e_2),$$

and by induction

$$J_0^k e_1 = b_1 b_2 \cdots b_k e_{k+1} + r_k, \tag{3.57}$$

where $r_k \in \text{Span}(e_1, \ldots, e_k)$. Thus,

$$e_1 \wedge J_0 e_1 \wedge \cdots \wedge J_0^k e_1$$
$$= e_1 \wedge (b_1 e_2 + r_1) \wedge (b_1 b_2 e_3 + r_2) \wedge \cdots \wedge (b_1 \ldots b_k e_{k+1} + r_k)$$
$$= b_1 (e_1 \wedge e_2 \wedge (b_1 b_2 e_3 + r_2) \wedge \cdots \wedge (b_1 \ldots b_k e_{k+1} + r_k))$$
$$= b_1 (b_1 b_2) (e_1 \wedge e_2 \wedge e_3 \wedge \cdots \wedge (b_1 \ldots b_k e_{k+1} + r_k))$$
$$= b_1 (b_1 b_2) \cdots (b_1 \ldots b_k) (e_1 \wedge e_2 \wedge \cdots \wedge e_{k+1}).$$

And therefore,

$$D_0 = 1, \quad D_k = b_1^2 (b_1 b_2)^2 \cdots (b_1 \ldots b_k)^2, \quad 1 \le k \le N-1. \tag{3.58}$$

Hence for $3 \le k \le N-1$, a straightforward telescopic argument shows that

$$\frac{D_k D_{k-2}}{D_{k-1}^2} = \frac{(b_1 \ldots b_k)^2}{(b_1 \ldots b_{k-1})^2} = b_k^2. \tag{3.59}$$

Also, for $k = 2$,

$$\frac{D_2 D_0}{D_1^2} = \frac{b_1^2 (b_1 b_2)^2}{b_1^4} = b_2^2,$$

and for $k = 1$,

$$\frac{D_1 D_{-1}}{D_0^2} = b_1^2,$$

where $D_{-1} := 1$; thus, with this convention,

$$b_k^2 = \frac{D_k D_{k-2}}{D_{k-1}^2}, \quad 1 \le k \le N-1. \tag{3.60}$$

Note that (3.60) expresses b_k, and hence $x_k - x_{k+1}$, directly in terms of the data $(\lambda_1, \ldots, \lambda_N)$ and $(u_1(1), \ldots, u_N(1)) = (u_{1,1}, \ldots, u_{1,N})$, whose evolution under the Toda flow is known explicitly from isospectrality and (3.23). We now express the b_k explicitly in terms of the λ_k and $u_k(1)$, so that the time evolution of the b_k may be read off.

Step 2: Expressing the b_k in terms of the spectral measure μ.

First, we expand D_k, $k \geq 1$, into a more useful formula, as follows: by (3.56) and the formula for a Vandermonde determinant,

$$D_k = \det \begin{pmatrix} \int d\mu(x_0) & \int x_1 d\mu(x_1) & \cdots & \int x_k^k d\mu(x_k) \\ \int x_0 d\mu(x_0) & \int x_1^2 d\mu(x_1) & & \int x_k^{k+1} d\mu(x_k) \\ \vdots & \vdots & & \vdots \\ \int x_0^k d\mu(x_0) & \int x_1^{k+1} d\mu(x_1) & \cdots & \int x_k^{2k} d\mu(x_k) \end{pmatrix}$$

$$= \int \cdots \int d\mu(x_0) \cdots d\mu(x_k) \det \begin{pmatrix} 1 & x_1 & \cdots & x_k^k \\ x_0 & x_1^2 & & x_k^{k+1} \\ \vdots & \vdots & & \vdots \\ x_0^k & x_1^{k+1} & \cdots & x_k^{2k} \end{pmatrix}$$

$$= \int \cdots \int d\mu(x_0) \cdots d\mu(x_k) x_0^0 x_1^1 \cdots x_k^k \det \begin{pmatrix} 1 & 1 & \cdots & 1 \\ x_0 & x_1 & & x_k \\ \vdots & \vdots & & \vdots \\ x_0^k & x_1^k & \cdots & x_k^k \end{pmatrix}$$

$$= \int \cdots \int d\mu(x_0) \cdots d\mu(x_k) x_0^0 x_1^1 \cdots x_k^k \prod_{0 \leq l < j \leq k} (x_j - x_l).$$

For a permutation σ of the set $\{0, 1, \ldots, k\}$, we find by averaging

$$D_k = \frac{1}{(k+1)!} \sum_\sigma \int d\mu(x_{\sigma(0)}) \cdots d\mu(x_{\sigma(k)}) x_{\sigma(0)}^0 \cdots x_{\sigma(k)}^k$$

$$\times \prod_{0 \leq l < j \leq k} (x_{\sigma(j)} - x_{\sigma(l)}),$$

where, for the parity $\epsilon(\sigma)$ of the permutation σ,

$$\prod_{0 \leq l < j \leq k} (x_{\sigma(j)} - x_{\sigma(l)}) = \epsilon(\sigma) \prod_{0 \leq l < j \leq k} (x_j - x_l).$$

Thus,

D_k

$$= \frac{1}{(k+1)!} \sum_\sigma \int d\mu(x_{\sigma(0)}) \cdots d\mu(x_{\sigma(k)}) x_{\sigma(0)}^0 \cdots x_{\sigma(k)}^k \epsilon(\sigma) \prod_{0 \leq l < j \leq k} (x_j - x_l)$$

$$= \frac{1}{(k+1)!} \int \cdots \int d\mu(x_0) \cdots d\mu(x_k) \left(\sum_\sigma \epsilon(\sigma) x_{\sigma(0)}^0 \cdots x_{\sigma(k)}^k \right) \prod_{0 \leq l < j \leq k} (x_j - x_l),$$

3.2 Long-Time Behavior

i.e., for $k \geq 1$,

$$D_k = \frac{1}{(k+1)!} \int \cdots \int d\mu(x_0) \cdots d\mu(x_k) \prod_{0 \leq l < j \leq k} (x_j - x_l)^2$$

$$= \int_{x_0 > x_1 > \cdots > x_k} d\mu(x_0) \cdots d\mu(x_k) \prod_{0 \leq l < j \leq k} (x_j - x_l)^2$$

as $\prod_{0 \leq l < j \leq k} (x_j - x_l)^2$ is invariant under permutations and terms with $x_i = x_{i+1}$ for some i just drop out. As before, $D_0 = 1$. Now, under the Toda flow,

$$d\mu(\lambda, t) = \sum_{i=1}^{N} u_i^2(1, t) \delta_{\lambda_i}(\lambda) = \frac{\sum_{i=1}^{N} u_i^2(1, 0) e^{2\lambda_i t} \delta_{\lambda_i}(\lambda)}{\sum_{i=1}^{N} u_i^2(1, 0) e^{2\lambda_i t}} = \frac{e^{2\lambda t} d\mu(\lambda)}{\int e^{2st} d\mu(s)},$$
(3.61)

where $d\mu(\lambda)$ is given by (3.56). Thus, under the Toda flow, for $k \geq 1$,

$$D_k(t) = \frac{1}{(\int e^{2xt} d\mu(x))^{k+1}} \int_{x_0 > \cdots > x_k} d\mu(x_0) \cdots d\mu(x_k) e^{2(x_0 + \cdots + x_k)t}$$
$$\times \prod_{0 \leq l < j \leq k} (x_j - x_l)^2,$$

and so from (3.60), for $k > 2$,

$$b_k^2 = \left(\int e^{2xt} d\mu(x) \right)^{-(k+1)-(k-1)+2k}$$
$$\times \int_{x_0 > \cdots > x_k} d\mu(x_0) \cdots d\mu(x_k) e^{2(x_0 + \cdots + x_k)t} \prod_{0 \leq l < j \leq k} (x_j - x_l)^2$$
$$\times \int_{x_0 > \cdots > x_{k-2}} d\mu(x_0) \cdots d\mu(x_{k-2}) e^{2(x_0 + \cdots + x_{k-2})t} \prod_{0 \leq l < j \leq k-2} (x_j - x_l)^2$$
$$\times \left(\int_{x_0 > \cdots > x_{k-1}} d\mu(x_0) \cdots d\mu(x_{k-1}) e^{2(x_0 + \cdots + x_{k-1})t} \prod_{0 \leq l < j \leq k-1} (x_j - x_l)^2 \right)^{-2}.$$

Thus, for $k > 2$,

$$b_k^2 = \int_{x_0 > \cdots > x_k} d\mu^k(x) e^{2(x_0 + \cdots + x_{k-2})t} V_k^2(x)$$
$$\times \int_{x_0 > \cdots > x_{k-2}} d\mu^{k-2}(x) e^{2(x_0 + \cdots + x_{k-2})t} V_{k-2}^2(x) \quad (3.62)$$
$$\times \left(\int_{x_0 > \cdots > x_{k-1}} d\mu^{k-1}(x) e^{2(x_0 + \cdots + x_{k-1})t} V_{k-1}^2(x) \right)^{-2},$$

where

$$d\mu^k(x) = d\mu(x_0)\ldots d\mu(x_k) \quad \text{and} \quad V_k(x) = \prod_{0 \le l < j \le k}(x_j - x_l). \quad (3.63)$$

For $k = 2$, from (3.60),

$$b_2^2 = \frac{D_2 D_0}{D_1^2}$$

$$= \left(\int e^{2xt}d\mu(x)\right)\frac{\int_{x_0 > x_1 > x_2} d\mu(x_0)d\mu(x_1)d\mu(x_2)e^{2(x_0+x_1+x_2)t}V_2^2(x)}{\left(\int_{x_0 > x_1} d\mu(x_0)d\mu(x_1)e^{2(x_0+x_1)t}V_1^2(x)\right)^2},$$

$$(3.64)$$

and for $k = 1$,

$$b_1^2 = \frac{1}{\left(\int e^{2xt}d\mu(x)\right)^2}\int_{x_0 > x_1} d\mu(x_0)d\mu(x_1)e^{2(x_0+x_1)t}V_1^2(x). \quad (3.65)$$

Step 3: Obtaining the asymptotic behavior of b_k as $t \to \pm\infty$.

Expanding (3.65), we obtain

$$b_1^2 = \frac{1}{2}\int_{\mathbb{R}^2} d\mu(x_0, t)d\mu(x_1, t)(x_1 - x_0)^2$$

$$= \frac{1}{2}\left(2\int d\mu(x_0, t)x_0^2 - 2\left(\int xd\mu(x, t)\right)^2\right) \quad (3.66)$$

$$= \int x^2 d\mu(x, t) - \left(\int xd\mu(x, t)\right)^2.$$

Recall from (3.29) that

$$b_1^2(t) = \sum_{i=1}^N \lambda_i^2 u_i^2(1, t) - \left(\sum_{i=1}^N \lambda_i u_i^2(1, t)\right)^2,$$

which matches (3.66), but now (3.65) yields the leading behavior of $b_1(t)$ directly. Indeed, as $t \to \infty$, noting that $V_k(x) = 0$ if $x_i = x_j$ for some $i \ne j$,

$b_1^2(t)$

$$= \frac{1}{\left(e^{2\lambda_1 t}u_1^2(1) + O(e^{2\lambda_2 t})\right)^2}(u_1^2(1)u_2^2(1)e^{2(\lambda_1+\lambda_2)t}(\lambda_1 - \lambda_2)^2 + o(e^{2(\lambda_1+\lambda_2)t}))$$

$$= \frac{u_2^2(1)}{u_1^2(1)}e^{-2(\lambda_1-\lambda_2)t}(\lambda_1 - \lambda_2)^2(1 + o(1)),$$

$$(3.67)$$

3.2 Long-Time Behavior

which agrees with (3.30). Closer examination of (3.65) shows that in fact the o(1) is exponentially small.

From now on, for $t \to \pm\infty$ we shall use the abbreviation e.s.e. to indicate exponentially small error terms of order $O(e^{-\gamma|t|})$ for some $\gamma > 0$.

Now, for $k > 2$, we have from (3.62), as $t \to \infty$,

$$b_k^2(t) = \left(u_1^2(1) \cdots u_{k+1}^2(1) e^{2(\lambda_1 + \cdots + \lambda_{k+1})t} \prod_{k \geq j > l \geq 0} (\lambda_{j+1} - \lambda_{l+1})^2 \right) (1 + e.s.e.)$$

$$\times \left(u_1^2(1) \cdots u_{k-1}^2(1) e^{2(\lambda_1 + \cdots + \lambda_{k-1})t} \prod_{k-2 \geq j > l \geq 0} (\lambda_{j+1} - \lambda_{l+1})^2 \right) (1 + e.s.e.)$$

$$\times \left(u_1^2(1) \cdots u_k^2(1) e^{2(\lambda_1 + \cdots + \lambda_k)t} \prod_{k-1 \geq j > l \geq 0} (\lambda_{j+1} - \lambda_{l+1})^2 \right)^{-2} (1 + e.s.e.).$$

Thus, for $3 \leq k \leq N - 1$,

$$b_k^2(t) = \frac{u_{k+1}^2(1)}{u_k^2(1)} e^{2(\lambda_{k+1} - \lambda_k)t} \frac{\prod_{l=0}^{k-1}(\lambda_{k+1} - \lambda_{l+1})^2}{\prod_{l=0}^{k-2}(\lambda_k - \lambda_{l+1})^2} (1 + e.s.e.). \qquad (3.68)$$

For $k = 2$, from (3.64) we have as $t \to \infty$,

$$b_2^2(t)$$
$$= (e^{2\lambda_1 t} u_1^2(1) + e.s.e.)$$
$$\times \frac{u_1^2(1) u_2^2(1) u_3^2(1) e^{2(\lambda_1 + \lambda_2 + \lambda_3)t} \left((\lambda_1 - \lambda_2)(\lambda_1 - \lambda_3)(\lambda_2 - \lambda_3)\right)^2}{\left(u_1^2(1) u_2^2(1) e^{2(\lambda_1 + \lambda_2)t}(\lambda_1 - \lambda_2)^2\right)^2} (1 + e.s.e.)$$
$$= \frac{u_3^2(1)}{u_2^2(1)} e^{2(\lambda_3 - \lambda_2)t} \frac{\prod_{l=0}^{1}(\lambda_3 - \lambda_{l+1})^2}{\prod_{l=0}^{k-2}(\lambda_2 - \lambda_{l+1})^2} (1 + e.s.e.).$$

Thus, we see that (3.68) also holds for $k \geq 2$, and by (3.67) also holds for $k = 1$, provided we use the convention

$$\prod_{l=0}^{-1} (\lambda_1 - \lambda_{l+1})^2 = 1. \qquad (3.69)$$

Exercise 3.12 Compute the analogous formulae for $a_k(t)$ when $t \to \infty$.

A similar calculation shows that, as $t \to -\infty$,

$$b_k^2(t) = \frac{u_{N-k}^2(1)}{u_{N-k+1}^2(1)} \times \frac{\prod_{l=N-k+1}^{N}(\lambda_{N-k} - \lambda_l)^2}{\prod_{l=N-k+2}^{N}(\lambda_{N-k+1} - \lambda_l)^2} e^{2t(\lambda_{N-k} - \lambda_{N-k+1})}(1 + e.s.e.). \qquad (3.70)$$

Alternatively, note that, under time reversal symmetry,

$$\tilde{x}_k(t) = x_k(-t), \quad \tilde{y}_k(t) = -y_k(-t), \quad 1 \le k \le N, \qquad (3.71)$$

also solve the Toda Lattice. Let $J(T)$ and $\tilde{J}(t)$ be the Jacobi operators associated with $(x_k(t), y_j(t))$ and $(\tilde{x}_k(t), \tilde{y}_j(t))$, respectively, as in (3.3) and (3.4). Let

$$\lambda_1 > \lambda_2 > \cdots > \lambda_N$$

be the eigenvalues of $J(0)$. Then as

$$J(0) = -\tilde{J}(0),$$

the eigenvalues of $\tilde{J}(0)$ are given by

$$\tilde{\lambda}_1 > \tilde{\lambda}_2 > \cdots > \tilde{\lambda}_N,$$

where

$$\tilde{\lambda}_k = -\lambda_{N-k+1}, \quad 1 \le k \le N.$$

As the eigenvectors of $J(0)$ and $\tilde{J}(0)$ are the same, the asymptotic behavior of $\tilde{b}_k(t), 1 \le k \le N$, as $t \to \infty$, can then be read off from (3.68) with $\{\lambda_k\}$ replaced by $\{\tilde{\lambda}_k\}$. But as $\tilde{x}_k(t) = x_k(-t)$, we immediately obtain the asymptotic behavior of $\tilde{b}_k(t), 1 \le k \le N$, as $t \to -\infty$. We leave the details to the interested reader.

An independent proof of (3.70) is also given in Chapter 4 using so-called Z variables: Here time reversal corresponds to a change of charts, one associated with the trivial permutation, the other to the permutation $\rho(i) = N+1-i, 1 \le i \le N$.

Again, formula (3.70) holds for $k = 1$, provided we interpret the denominator in (3.70) as 1, i.e.,

$$b_1^2(t) = \frac{u_{N-1}^2(1)}{u_N^2(1)}(\lambda_{N-1} - \lambda_N)^2 e^{2t(\lambda_{N-1} - \lambda_N)}(1 + \text{e.s.e.}).$$

Step 4: Asymptotic behavior of x_k as $t \to \pm\infty$.

We now convert the asymptotic formulae for the $b_k(t)$ into asymptotic formulae for $x_k(t)$ as $t \to \pm\infty$. From (3.68) and (3.70) we have, as $t \to +\infty$,

$$\begin{aligned}x_k - x_{k+1} &= 2\ln 2 + 2\ln b_k \\ &= 2t(\lambda_{k+1} - \lambda_k) + 2\ln\left(2\frac{u_{k+1}(1)}{u_k(1)}\frac{\prod_{l=1}^{k}(\lambda_l - \lambda_{k+1})}{\prod_{l=1}^{k-1}(\lambda_l - \lambda_k)}\right) + \text{e.s.e.}\end{aligned}$$

$$(3.72)$$

and as $t \to -\infty$,

$$x_k - x_{k+1} = 2t(\lambda_{N-k} - \lambda_{N-k+1}) \tag{3.73}$$

$$+ 2\ln\left(2\frac{u_{N-k}(1)}{u_{N-k+1}(1)} \times \frac{\prod_{l=N-k+1}^{N}(\lambda_{N-k} - \lambda_l)}{\prod_{l=N-k+2}^{N}(\lambda_{N-k+1} - \lambda_l)}\right) + \text{e.s.e.} \tag{3.74}$$

Now

$$\partial_t \sum_{k=1}^{N} x_k = \sum_{k=1}^{N} y_k = -2\sum_{k=1}^{N} a_k = -2\sum_{i=1}^{N} \lambda_i = \text{const.}$$

and so

$$\sum_{k=1}^{N} x_k(t) = \sum_{k=1}^{N} x_k(0) - 2\left(\sum_{i=1}^{N} \lambda_i\right) t. \tag{3.75}$$

Summing (3.72), we obtain, as $t \to \infty$, for $1 \le k \le N-1$,

$$x_k - x_N = 2t(\lambda_N - \lambda_k) + (\ln 4)(N-k) + 2\ln\left(\frac{u_N(1)}{u_k(1)} \frac{\prod_{l=1}^{N-1}(\lambda_l - \lambda_N)}{\prod_{l=1}^{k-1}(\lambda_l - \lambda_k)}\right) + \text{e.s.e.}, \tag{3.76}$$

and so, summing over k,

$$\sum_{k=1}^{N-1} x_k - (N-1)x_N = 2t\left((N-1)\lambda_N - \sum_{k=1}^{N-1} \lambda_k\right) + \ln 4 \sum_{k=1}^{N-1}(N-k)$$

$$+ 2\sum_{k=1}^{N-1} \ln\left(\frac{u_N(1)}{u_k(1)} \frac{\prod_{l=1}^{N-1}(\lambda_l - \lambda_N)}{\prod_{l=1}^{k-1}(\lambda_l - \lambda_k)}\right) + \text{e.s.e.}$$

Inserting (3.75), we find that as $t \to \infty$,

$$x_N(t) = \frac{1}{N}\sum_{k=1}^{N} x_k(0) - 2t\lambda_N - \frac{\ln 4}{N}\sum_{k=1}^{N-1}(N-k)$$

$$- \frac{2}{N}\sum_{k=1}^{N-1} \ln\left(\frac{u_N(1)}{u_k(1)} \frac{\prod_{l=1}^{N-1}(\lambda_l - \lambda_N)}{\prod_{l=1}^{k-1}(\lambda_l - \lambda_k)}\right) + \text{e.s.e.},$$

which implies using (3.76) that, as $t \to \infty$,

$$x_k(t) = -2t\lambda_k + \frac{1}{N}\sum_{j=1}^{N} x_j(0) - \frac{\ln 4}{N}\sum_{j=1}^{N-1}(N-j) + (N-k)\ln 4$$

$$- \frac{2}{N}\sum_{j=1}^{N} \ln\left(\frac{u_N(1)}{u_j(1)} \frac{\prod_{l=1}^{N-1}(\lambda_l - \lambda_N)}{\prod_{l=1}^{j-1}(\lambda_l - \lambda_j)}\right)$$

$$+ 2\ln\left(\frac{u_N(1)}{u_k(1)} \frac{\prod_{l=1}^{N-1}(\lambda_l - \lambda_N)}{\prod_{l=1}^{k-1}(\lambda_l - \lambda_k)}\right) + \text{e.s.e.}$$

After some elementary algebra, this reduces to

$$x_k(t) = -2t\lambda_k + \frac{1}{N}\sum_{j=1}^{N} x_j(0) - \frac{2}{N}\sum_{j=1}^{N} \ln\left(\frac{u_k(1)}{u_j(1)} \frac{\prod_{l=1}^{k-1} 2(\lambda_l - \lambda_k)}{\prod_{l=1}^{j-1} 2(\lambda_l - \lambda_j)}\right) + \text{e.s.e.} \tag{3.77}$$

as $t \to \infty$, and a similar calculation using (3.73) yields

$$x_k(t) = -2t\lambda_{N-k+1} + \frac{1}{N}\sum_{j=1}^{N} x_j(0)$$

$$- \frac{2}{N}\sum_{j=1}^{N} \ln\left(\frac{u_{N-k+1}(1)}{u_{N-j+1}(1)} \frac{\prod_{l=N-k+2}^{N} 2(\lambda_l - \lambda_{N-k+1})}{\prod_{l=N-j+2}^{N} 2(\lambda_l - \lambda_{N-j+1})}\right) + \text{e.s.e.}$$

as $t \to -\infty$. Summarizing, we have shown that for $1 \le k \le N$,

$$x_k(t) = -2t\lambda_k + \beta_k^+ + o(1) \qquad \text{as } t \to +\infty, \tag{3.78}$$

and

$$x_k(t) = -2t\lambda_{N-k+1} + \beta_k^- + o(1) \qquad \text{as } t \to -\infty, \tag{3.79}$$

where

$$\beta_k^+ = \frac{1}{N}\sum_{j=1}^{N} x_j(0) - \frac{2}{N}\sum_{j=1}^{N} \ln\left(\frac{u_k(1)}{u_j(1)} \frac{\prod_{l=1}^{k-1} 2(\lambda_l - \lambda_k)}{\prod_{l=1}^{j-1} 2(\lambda_l - \lambda_j)}\right), \tag{3.80}$$

and

$$\beta_k^- = \frac{1}{N}\sum_{j=1}^{N} x_j(0) - \frac{2}{N}\sum_{j=1}^{N} \ln\left(\frac{u_{N-k+1}(1)}{u_{N-j+1}(1)} \frac{\prod_{l=N-k+2}^{N} 2(\lambda_l - \lambda_{N-k+1})}{\prod_{l=N-j+2}^{N} 2(\lambda_l - \lambda_{N-j+1})}\right). \tag{3.81}$$

3.2 Long-Time Behavior

Step 5: Computing the phase shifts $\beta_k^+ - \beta_{N-k+1}^-$.

We have

$$\beta_k^+ - \beta_{N-k+1}^-$$

$$= -\frac{2}{N} \sum_{j=1}^{N} \ln\left(\frac{u_k(1) \prod_{l=1}^{k-1}(2\lambda_l - 2\lambda_k) \, u_j(1) \prod_{l=j+1}^{N}(2\lambda_l - 2\lambda_j)}{u_j(1) \prod_{l=1}^{j-1}(2\lambda_l - 2\lambda_j) \, u_k(1) \prod_{l=k+1}^{N}(2\lambda_l - 2\lambda_k)}\right)$$

$$= -\frac{1}{N} \sum_{j=1}^{N} \left(\sum_{l=1}^{k-1} \ln(2\lambda_l - 2\lambda_k)^2 - \sum_{l=k+1}^{N} \ln(2\lambda_k - 2\lambda_l)^2 \right)$$

$$- \frac{1}{N} \sum_{j=1}^{N} \left(\sum_{l=j+1}^{N} \ln(2\lambda_j - 2\lambda_l)^2 - \sum_{l=1}^{j-1} \ln(2\lambda_l - 2\lambda_j)^2 \right).$$

Thus

$$\beta_k^+ - \beta_{N-k+1}^- = \sum_{l \neq k} \phi_{lk}, \qquad 1 \leq k \leq N, \tag{3.82}$$

where

$$\phi_{lk} = \begin{cases} \ln(2\lambda_l - 2\lambda_k)^2, & l > k, \\ -\ln(2\lambda_l - 2\lambda_k)^2, & l < k. \end{cases} \tag{3.83}$$

Here we have used the convention (cf. (3.69)) that

$$\sum_{l=N+1}^{N} \ln(2\lambda_N - 2\lambda_l)^2 = 0 = \sum_{l=1}^{0} \ln(2\lambda_l - 2\lambda_1)^2,$$

and so

$$\sum_{j=1}^{N} \sum_{l=j+1}^{N} \ln(2\lambda_j - 2\lambda_l)^2 = \sum_{1 \leq j < l \leq N} \ln(2\lambda_j - 2\lambda_l)^2$$

$$= \sum_{j=2}^{N} \sum_{l=1}^{j-1} \ln(2\lambda_l - 2\lambda_j)^2$$

$$= \sum_{j=1}^{N} \sum_{l=1}^{j-1} \ln(2\lambda_l - 2\lambda_j)^2.$$

In particular, we see that for $N = 2$,

$$\beta_1^+ - \beta_2^- = \ln(2\lambda_1 - 2\lambda_2)^2,$$
$$\beta_2^+ - \beta_1^- = -\ln(2\lambda_1 - 2\lambda_2)^2.$$

Thus the phase shifts

$$\beta_k^+ - \beta_{N-k+1}^- = \sum_{j \neq k} \phi_{jk},$$

are just the same as if the interaction takes place two particles at a time.

Remark 3.13 The remarkable formula (3.82) for the phase shift is due to Moser and Zehnder (2005), which they derived using an indirect, and very different, argument.

Exercise: Note that the phase shifts (3.82) depend only on the eigenvalues of $L(0)$. This shows that the phase shifts are constants of the motion for the Toda flow. Give a dynamical proof of this fact.

Exercise 3.14 Use (3.60) and (3.61) to show that

$$x_j(t) = x_1(t) + \ln\left(\frac{D_{j-2}(t)}{D_{j-1}(t)}\right) - (j-1)\ln 4, \qquad (3.84)$$

where

$$x_1(t) = \frac{1}{N}\sum_{i=1}^N x_i(0) - 2\frac{t}{N}\sum_{i=1}^N \lambda_i + (N-2)\ln 2 + \frac{1}{N}\ln D_{N-1}(t)$$

$$= \frac{1}{N}\sum_{i=1}^N x_i(0) - 2\frac{t}{N}\sum_{i=1}^N \lambda_i + (N-2)\ln 2$$

$$+ \frac{1}{N}\ln\left(V^2(\lambda)\frac{e^{2\sum_{i=1}^N \lambda_i t}}{\sum_{i=1}^N e^{2\lambda_i t}u_i^2(1)}\prod_{i=1}^N u_i^2(1)\right).$$

Use the above equations to rederive (3.78) and (3.79).

3.3 Liouville Integrability of the Toda Lattice

We show that the Toda lattice is integrable in the sense of Liouville, i.e.,

$$H_{\text{Toda}}(x, y) = \frac{1}{2}\sum_{i=1}^N y_i^2 + \sum_{i=1}^{N-1} e^{x_i - x_{i+1}}$$

has N independent Poisson-commuting integrals $\{I_j\}$ on the symplectic manifold $(\mathbb{R}^{2N}, \omega = \sum dx_i \wedge dy_i)$.

We begin with an extension of Theorem 1.1. For a polynomial $p : \mathbb{R} \to \mathbb{R}$, consider the *generalized Toda equation*

$$\partial_t X = [X, B(p(X))] = XB(p(X)) - B(p(X))X, \qquad X(0) = X_0, \qquad (3.85)$$

where

3.3 Liouville Integrability of the Toda Lattice

$$B(p(X)) = p(X)_- - p(X)_-^T, \tag{3.86}$$

$p(X)_-$ being the strictly lower triangular part of $p(X)$.

Theorem 3.15 *The solution $X(t)$ of the generalized Toda equation is globally defined and admits two representations by conjugation,*

$$X(t) = Q(t)^T X_0 Q(t) \quad \text{and} \quad X(t) = R(t) X_0 R(t)^{-1},$$

where $Q(t)$ are orthogonal matrices and $R(t)$ are upper triangular matrices with positive diagonal entries and $Q(0) = R(0) = 0$. If X_0 is real symmetric the solution $X(t)$ is also real symmetric. If X_0 is Jacobi, so is $X(t)$.

Proof To obtain global existence for the solution and the conjugation formulas, simply imitate the proof of Theorem 1.1. To show the preservation of symmetry and of Jacobi form, we follow the argument in Remark 3.2: Symmetry is preserved under orthogonal conjugation and conjugation by upper triangular matrices preserves the zeros at entries $X_{ij}, i > j + 1$. □

Equation (3.10) shows that the Toda flow is induced by the Hamiltonian $H_{\text{Toda}}(x, y) = 2 \, \text{tr} \, J^2(a(x, y), b(x, y))$. We now show that generalized Toda flows are also Hamiltonian.

Theorem 3.16 *Let P be a primitive of the polynomial p. Then*

$$H(x, y) = 4 \, \text{tr} \, P(J(a(x, y), b(x, y)))$$

is a Hamiltonian inducing the generalized Toda flow on Jacobi matrices,

$$\partial_t J = [J, B(p(J))], \qquad J(0) = J_0. \tag{3.87}$$

Proof For a Hamiltonian $H_k(x, y) = \frac{4}{k} \, \text{tr} \, J^k$, since $\text{tr} \, AB = \text{tr} \, BA$, we have

$$\partial_t H_k = \frac{4}{k} \sum_{i=0}^{k-1} \text{tr} \, J^i \, \partial_t J \, J^{k-i-1} = 4 \, \text{tr} \, J^{k-1} \partial_t J.$$

By linearity, for $H(J) = 4 \, \text{tr} \, P(J)$,

$$\partial_t H = 4 \, \text{tr} \, p(J) \, \partial_t J.$$

We follow the proof of Theorem 3.1. Let $b_0 = b_N = 0$. For $1 \leq i \leq N$,

$$\partial_t a_i = \sum_{k=1}^N \partial_{x_k} a_i \, \partial_{y_k} H - \partial_{y_k} a_i \, \partial_{x_k} H = \frac{1}{2} \partial_{x_i} H = 2 \, \text{tr} \, p(J) \, \partial_{x_i} J$$

$$= \text{tr} \, p(J) \, (-b_{i-1} E_{i,i-1} - b_{i-1} E_{i-1,i} + b_i E_{i+1,i} + b_i E_{i,i+1})$$

$$= 2 \, \text{tr} \, p(J) \, (-b_{i-1} E_{i,i-1} + b_i E_{i+1,i})$$

$$= -2 b_{i-1} p(J)_{i-1,i} + 2 b_i p(J)_{i+1,i}, \tag{3.88}$$

where $E_{ij} = e_i e_j^T$ is the matrix with a single nonzero entry (i, j) equal to 1. For $1 \leq i \leq N - 1$,

$$\partial_t b_i = \sum_{k=1}^{N} \partial_{x_k} b_i \, \partial_{y_k} H - \partial_{y_k} b_i \, \partial_{x_k} H = \sum_{k=1}^{N} \partial_{x_k} b_i \, \partial_{y_k} H$$

$$= \partial_{x_i} b_i \, \partial_{y_i} H + \partial_{x_{i+1}} b_i \, \partial_{y_{i+1}} H = \frac{1}{2}(b_i \, \partial_{y_i} H - b_i \, \partial_{y_{i+1}} H)$$

$$= 2b_i \left(\operatorname{tr} \, p(J) \, (\partial_{y_i} J - \partial_{y_{i+1}} J) \right) = b_i \left(\operatorname{tr} \, p(J) \, (-E_{ii} + E_{i+1,i+1}) \right)$$

$$= b_i(-p(J)_{ii} + p(J)_{i+1,i+1}). \tag{3.89}$$

Write $B = B(p(J))$. We compare these equations with the evolutions for the entries of J given by the commutator in Equation (3.85),

$$\partial_t J = [J, B] = JB - BJ = JB + (JB)^T. \tag{3.90}$$

By Theorem 3.15, evolutions preserve the Jacobi form of the initial condition: We must verify equality at diagonal and subdiagonal entries (ii) and $(i + 1, i)$. Split $p(J) = p(J)_- + p(J)_0 + p(J)_+$ in strictly lower, diagonal, and upper triangular parts. As J is tridiagonal and the diagonal of B is zero, we have

$$\partial_t J_{ii} = [J, B]_{ii} = 2(JB)_{ii} = 2(e_i^T J)(Be_i) = 2b_{i-1} B_{i-1,i} + 2b_i B_{i+1,i},$$

$$\partial_t J_{i+1,i} = [J, B]_{i+1,i} = [J, p(J) - p(J)_0 - 2p(J)_+]_{i+1,i}$$

$$= [J, -p(J)_0 - 2p(J)_+]_{i+1,i} = -[J, p(J)_0]_{i+1,i} - 2[J, p(J)_+]_{i+1,i}$$

$$= b_i(-p(J)_{ii} + p(J)_{i+1,i+1}). \tag{3.91}$$

In the last step we used the fact that the commutator of a Jacobi matrix with a strictly upper triangular matrix is upper triangular.

From the definition of $B = B(p(J))$ in terms of $p(J)$, we have

$$B_{i+1,i} = p(J)_{i+1,i}, \quad B_{i,i+1} = -p(J)_{i,i+1}$$

and thus

$$\begin{aligned} \partial_t J_{ii} &= -2b_{i-1} p(J)_{i-1,i} + 2b_i p(J)_{i+1,i}, \\ \partial_t J_{i+1,i} &= b_i(-p(J)_{ii} + p(J)_{i+1,i+1}). \end{aligned} \tag{3.92}$$

The results agree with equations (3.88) and (3.89). □

We take $I_j = \lambda_j$, the eigenvalues of J_0, $1 \leq j \leq N$. We already know these are integrals for the Toda flow: It remains to show that they Poisson-commute and are independent.

3.4 The Toda Lattice: Full Matrix Case

Theorem 3.17 *Let $\{\lambda_j, 1 \leq j \leq N\}$ be the eigenvalues of an $N \times N$ Jacobi matrix J. Then*

$$\{\lambda_k, \lambda_j\} = 0, \quad 1 \leq k, j \leq N. \tag{3.93}$$

Proof From Theorem 3.16, for $q = 1, \ldots, N$, the Hamiltonian $H_q(x, y) = \frac{4}{q} \operatorname{tr} J^q$, induces a flow in Lax pair form and hence the eigenvalues of J_0 are conserved under the flow. Thus

$$\{\lambda_j, H_q\} = \partial_t \lambda_j = 0, \quad 1 \leq j, q \leq N.$$

Hence, for any $p = 1, \ldots, N$,

$$\{H_p, H_q\} = 4/p \sum_j \{\lambda_j^p, H_q\} = 4 \sum_j \lambda_j^{p-1} \{\lambda_j, H_q\} = 0.$$

But then $0 = \sum_{i,j} \{\lambda_i^p, \lambda_j^q\} = \sum_{i,j} pq\, \lambda_i^{p-1} \lambda_j^{q-1} \{\lambda_i, \lambda_j\}$. As the eigenvalues $\{\lambda_i\}$ are distinct, the Vandermonde determinant $\{\lambda_i^{p-1}\}_{1 \leq i, p \leq N}$ is nonzero and hence

$$\sum_j \lambda_j^{q-1} \{\lambda_i, \lambda_j\} = 0, \quad 1 \leq i, q \leq N.$$

Again using the Vandermonde determinant, we conclude that

$$\{\lambda_i, \lambda_j\} = 0, \quad 1 \leq i, j \leq N. \qquad \square$$

Theorem 3.18 *Let $\{\lambda_j, 1 \leq j \leq N\}$ be the eigenvalues of an $N \times N$ Jacobi matrix X. Then, $\lambda_1, \ldots, \lambda_N$ are independent.*

Proof By Theorem 3.8, the map $\psi : \mathcal{J} \to M, J \mapsto (\lambda_1, \ldots, \lambda_N, u_1, \ldots, u_N)$ is a diffeomorphism. In particular, the first N columns of its Jacobian (the matrix of first-order derivatives) at any point must be independent. $\square$

We will construct the *angles* corresponding to these *actions*, viz. the eigenvalues $\lambda_1, \ldots, \lambda_N$, in Section 3.8. The existence of such *action-angles* variables is guaranteed by the Liouville–Arnold–Jost theorem 2.13.

3.4 The Toda Lattice: Full Matrix Case

In this section we consider the generalized Toda flow

$$\partial_t X = [X, B(X)]$$

on full real symmetric matrices $X \in \Sigma_N$. A general reference for what follows is Deift et al. (1986): see also Symes (1982). Consider the group Lo of lower triangular $N \times N$ matrices

$$Lo := \{L \ : \ L_{ij} = 0 \text{ for } 1 \le i < j \le N, L_{ii} > 0\}.$$

The Lie algebra $\mathfrak{lo}$ of Lo also consists of lower triangular matrices:

$$\mathfrak{lo} = \{M \ : \ M_{ij} = 0 \text{ for } 1 \le i < j \le N\}.$$

The Lie bracket for $\mathfrak{lo}$ is just the matrix commutator

$$[M_1, M_2] = M_1 M_2 - M_2 M_1$$

for $M_1, M_2 \in \mathfrak{lo}$. The dual Lie algebra $\mathfrak{lo}^*$ of Lo can clearly be identified with the $N \times N$ real symmetric matrices Σ_N via the nondegenerate pairing

$$\lambda_X(M) := \langle X, M \rangle = \text{tr}(XM), \quad X \in \Sigma_N, M \in \mathfrak{lo}. \tag{3.94}$$

As described in Chapter 2, Lo acts on $\mathfrak{lo}$ via the Ad-action. (Here, the group operation is just matrix multiplication.)

$$Ad_{Lo}(M) = \partial_t L e^{tM} L^{-1}|_{t=0} = LML^{-1} \in \mathfrak{lo}. \tag{3.95}$$

For $L \in Lo$, $M \in \mathfrak{lo}$. Lo acts on $\mathfrak{lo}^*$ via the Ad^* action

$$\langle Ad_L^* X, M \rangle = \langle X, Ad_L M \rangle = \langle X, LML^{-1} \rangle$$
$$= \text{tr}\, XLML^{-1} = \text{tr}(L^{-1}XL)M.$$

Now, for any $N \times N$ matrix Y, set

$$\pi_{k\perp} Y := Y_+ + Y_0 + Y_+^T, \tag{3.96}$$

and

$$\pi_{\mathfrak{lo}\perp} Y := Y_- - Y_+^T, \tag{3.97}$$

where Y_+, Y_- are the strictly upper (resp. lower) triangular parts of Y, and $Y_0 = \text{diag}(Y)$. Clearly,

$$\begin{aligned} \pi_{k\perp} + \pi_{\mathfrak{lo}\perp} &= \text{Id}, \\ \pi_{k\perp} Y \in \mathfrak{lo}^*, &\quad \pi_{\mathfrak{lo}\perp} Y \text{ is strictly lower triangular.} \end{aligned} \tag{3.98}$$

Now, as $\text{tr}(\pi_{\mathfrak{lo}\perp} Y)M = 0$ for any Y and any $M \in \mathfrak{lo}$, we conclude that

$$\langle Ad_L^* X, M \rangle = \text{tr}(\pi_{k\perp} L^{-1} XL) M,$$

and we obtain an explicit formula for the coadjoint action,

$$Ad_L^* X = \pi_{k\perp} L^{-1} XL. \tag{3.99}$$

3.4 The Toda Lattice: Full Matrix Case

Let $\mathcal{O}_X$ denote the coadjoint orbit through a point $X \in \mathfrak{lo}^*$,

$$\mathcal{O}_X = \{\mathrm{Ad}_L^* X\}, \quad L \in Lo. \tag{3.100}$$

Then, by general theory, as noted in Section 2.2, $\mathcal{O}_X$ is an even dimensional symplectic manifold with some nondegenerate 2-form ω_X, the Kirillov–Kostant symplectic structure (Kirillov, 2004). Also, $\mathfrak{lo}^*$ carries a (degenerate) Poisson structure: for smooth functions $H, K : \mathfrak{lo}^* \mapsto \mathbb{R}$,

$$\{H, K\}(X) = \langle X, [dH(X), dK(X)] \rangle = \mathrm{tr}\, \hat{X}[dH(\hat{X}), dK(\hat{X})]. \tag{3.101}$$

Here, $dH(X)$ is the linear functional on $\mathfrak{lo}^*$ given by

$$dH(X)(\hat{X}) = \partial_t H(X + t\hat{X})|_{t=0} \tag{3.102}$$

for $\hat{X} \in \mathfrak{lo}^*$, and as $\mathfrak{lo}^{**} = \mathfrak{lo}$, $dH(X) \in \mathfrak{lo}$. Similarly, $dK(X) \in \mathfrak{lo}$ and so $[dH(X), dK(X)] \in \mathfrak{lo}$.

The Poisson structure on $\mathfrak{lo}^*$ and the 2-form ω_X on $\mathcal{O}_X$ are related as follows (cf. Exercise 2.1):

$$\omega_X(v_H(X'), v_K(X')) = \{H, K\}(X')$$

for $X' \in \mathcal{O}_X$ and for all Hamiltonians H, K on $\mathcal{O}_X$.

In what follows, for convenience, we will always assume that $H : \mathfrak{lo}^* \to \mathbb{R}$ is the restriction to $\mathfrak{lo}^*$ of a function $\tilde{H} : \mathcal{M}_N(\mathbb{R}) \to \mathbb{R}$. This assumption involves no loss of generality: indeed, given $H : \mathfrak{lo}^* \to \mathbb{R}$, $\tilde{H}(Y) := H((Y + Y^T)/2)$ maps $\mathcal{M}_N(\mathbb{R}) \to \mathbb{R}$ and $\tilde{H}(X) = H(X)$ whenever $X \in \mathfrak{lo}^*$. Set

$$\pi_k(X) = X_+ - X_+^T, \quad \pi_l(X) = X_- + X_0 + X_+^T. \tag{3.103}$$

Clearly $\pi_k(X) + \pi_l(X) = X$.

Theorem 3.19 *For smooth functions $H, K : \mathfrak{lo}^* \to \mathbb{R}$ and $X \in \mathfrak{lo}^*$, we have*

$$v_H(X) = \pi_{\mathfrak{lo}} \nabla H(X), \quad v_K(X) = \pi_{\mathfrak{lo}} \nabla H(K). \tag{3.104}$$

Denote by v_H, v_K the vector fields on $\mathcal{O}_X$ induced by the restrictions of H and K to $\mathcal{O}_X$. For $\hat{X} \in \mathcal{O}_X$,

$$\omega_X(\hat{X})(v_H(\hat{X}), v_K(\hat{H})) = \mathrm{tr}\, \hat{X}[dH(\hat{X}), dK(\hat{X})] = \mathrm{tr}\, \hat{X}[\pi_{\mathfrak{lo}} \nabla H(\hat{X}), \pi_{\mathfrak{lo}} \nabla K(\hat{X})]. \tag{3.105}$$

Let $X(t)$ be the flow generated by the Hamiltonian $H_q(X) = \frac{1}{q}\mathrm{tr}\, X^q$ on $(\mathcal{O}_{X_0}, \omega_{X_0})$ for any (not necessarily tridiagonal) $X_0 \in \mathfrak{lo}^*$. Then, for $X \in \mathcal{O}_{X_0}$,

$$\partial_t X = [X, \pi_k X^{q-1}]. \tag{3.106}$$

Proof To evaluate dH, note that, for $1 \le i, j \le N$,

$$dH(X)(\hat{X}) = \sum_{i,j} \left(\partial_{X_{ij}} H\right)(\hat{X}_{ij}) = \operatorname{tr} \nabla H \hat{X},$$

as $\hat{X} = \hat{X}^T$, where $\nabla H := (\partial_{X_{ij}} H)$. Now, $\nabla H = \pi_k \nabla H + \pi_{\mathfrak{l}_0} \nabla H$. But, as $\pi_k \nabla H$ is skew, $\operatorname{tr}(\pi_k \nabla H)\hat{X} = 0$. Hence, $dH(X)(\hat{X}) = \operatorname{tr}(\pi_{\mathfrak{l}_0} \nabla H)\hat{X}$ and $\upsilon_H(\hat{X}) = \operatorname{tr}(\pi_{\mathfrak{l}_0} \nabla K(\hat{X}))$. Moreover,

$$\omega_X(\hat{X})(\upsilon_H(\hat{X}), \upsilon_K(\hat{H})) = \{H, K\}_X(\hat{X})$$
$$= \langle \hat{X}, [dH(\hat{X}), dK(\hat{X})] \rangle = \operatorname{tr} \hat{X}[\pi_{\mathfrak{l}_0} \nabla H(\hat{X}), \pi_{\mathfrak{l}_0} \nabla K(\hat{X})].$$

Now consider the Hamiltonian on $\mathcal{O}_{X_0}$ given by

$$H_q(X) = \frac{1}{q} \operatorname{tr} X^q, \quad q = 1 \ldots N. \tag{3.107}$$

For the associated flow $X(t)$, for any function $K : \mathfrak{l}_0^* \to \mathbb{R}$, we have from (2.3),

$$\partial_t K = \{K, H_q\}_{X_0} = \operatorname{tr}(X[\pi_l \nabla K, \pi_l \nabla H_q]). \tag{3.108}$$

We compute the gradient:

$$(\nabla H_q(X))_{ij} = \operatorname{tr} X^{q-1} e_i e_j^T = \langle e_j, X^{q-1} e_i \rangle = (X^{q-1})_{ji}.$$

Hence, as $X = X^T$,

$$\nabla H_q(X) = X^{q-1}, \tag{3.109}$$

$$\partial_t K(X_t) = \sum_{ij}(\partial_{X_{ij}} K)(\partial_t X_{ij}) = \operatorname{tr}((\partial_t X)(\nabla K)).$$

Now, if Y is symmetric and S is skew,

$$\operatorname{tr} YS = \operatorname{tr} S^T Y^T = -\operatorname{tr} SY = -\operatorname{tr} YS,$$

and hence $\operatorname{tr} YS = 0$. But $\nabla K = \pi_k \nabla K + \pi_l \nabla K$, and as $\pi_k \nabla K$ is skew,

$$\partial_t K(X(t)) = \operatorname{tr}(\partial_t X \pi_l \nabla K). \tag{3.110}$$

However, from (3.108),

$$\partial_t K = \operatorname{tr}(X[\pi_l \nabla K, \pi_l \nabla H_q]).$$

But $\operatorname{tr}(A[B,C]) = \operatorname{tr}(ABC - ACB) = \operatorname{tr}(CAB - ACB) = \operatorname{tr}([C,A]B)$ and so

$$\partial_t K(X(t)) = \operatorname{tr}([\pi_l \nabla H_q(X), X]\pi_l \nabla K) = \operatorname{tr}((\pi_{k^T}[\pi_l \nabla H_q(X), X])\pi_l \nabla K) \tag{3.111}$$

as $\pi_{kT} + \pi_{lT} = \text{Id}$ and $\text{tr}(\pi_{lT}A)(\pi_l B) = 0$ for any A, B. Comparing (3.110) and (3.111), it follows that as K is arbitrary,

$$\partial_t X = \pi_{kT}[\pi_l(\nabla H_q(X)), X] = \pi_{kT}([\pi_l X^{q-1}, X]), \qquad (3.112)$$

where we have also used (3.109). Write

$$\pi_l X^{q-1} = X^{q-1} - \pi_k X^{q-1},$$

and, as $[X^{q-1}, X] = 0$,

$$\partial_t X = \pi_{kT}[X, \pi_k X^{q-1}].$$

As X is symmetric and $\pi_k X^{q-1}$ is skew, $[X, \pi_k X^{q-1}]$ is symmetric. But $\pi_{kT} A = A$ for any symmetric matrix A, and finally we conclude that

$$\partial_t X = [X, \pi_k X^{q-1}].$$

$\square$

We also consider more general Hamiltonians $H_f(X) = \text{tr} f(X)$, where $f : U \to \mathbb{C}$ is a smooth function on U, an open set containing $\sigma(X)$, with f real valued on $U \cap \mathbb{R}$. Here $f(X)$ is given by the functional calculus for real, symmetric matrices X: For a spectral decomposition $X = O^T DO$, set $f(X) = O^T f(D) O$. There is a problem in this approach, however, in that the map $X \mapsto (O, D)$ is not smooth in general, and so the smoothness of f does not necessarily imply the smoothness of the map $X \mapsto f(X)$; thus $H_f(X) = \text{tr} f(X)$ so defined is not necessarily C^1. In the case where f is analytic in U, however, this difficulty is overcome using the Dunford integral. Recall that for an analytic function $f : U \subset \mathbb{C} \to \mathbb{C}$ and X a real, symmetric matrix with $\sigma(X) \subset U$, the Dunford integral (Dunford and Schwartz, 1988) is given by

$$f(X) := \frac{1}{2i\pi} \int_{\mathcal{C}} \frac{f(s)}{s - X} ds, \qquad (3.113)$$

where $\mathcal{C}$ is a simple, closed positively oriented contour in U enclosing $\sigma(X)$ in its interior. Clearly $f(X)$ is a smooth function of X. Note that by the spectral theorem and Cauchy's integral formula, the integral in (3.113) produces a matrix of the form $O^T f(D) O$, where O is some real orthogonal matrix. It follows in particular that if f is real-valued on $U \cap \mathbb{R}$, then $f(X)$ defined by (3.113) is a real symmetric matrix.

Theorem 3.20 *For $X \in \mathfrak{lo}^*$ and $f : U \to \mathbb{C}$ as above with f real-valued* on $U \cap \mathbb{R}$, define $f(X)$ by the Dunford integral (3.113). Then $H_f(X) = \text{tr } f(X)$ is a smooth real-valued function which generates the vector field

$$\partial_t X = [X, \pi_k f'(X)], \qquad (3.114)$$

where f' is the derivative of f.

Proof From equation (3.113), we have

$$\begin{aligned}
\frac{\partial}{\partial X_{ij}} \text{tr } f(X) &= \frac{1}{2i\pi} \int_{\mathcal{C}} f(s) \text{ tr}\left(\frac{1}{s-X} e_i e_j^T \frac{1}{s-X}\right) ds \\
&= \frac{1}{2i\pi} \int_{\mathcal{C}} f(s) \left\langle e_j, \frac{1}{(s-X)^2} e_i \right\rangle ds \\
&= -\frac{1}{2i\pi} \int_{\mathcal{C}} f(s) \frac{d}{ds}\left\langle e_j, \frac{1}{(s-X)} e_i \right\rangle ds \\
&= \frac{1}{2i\pi} \int_{\mathcal{C}} f'(s) \left\langle e_j, \frac{1}{(s-X)} e_i \right\rangle ds \\
&= (f'(X))_{ji} = (f'(X))_{ij}, \qquad (3.115)
\end{aligned}$$

where $f'(X)$ is defined as in (3.113) with f replaced by f'. Using

$$\nabla H(X) = \nabla \text{tr } f(X) = f'(X)$$

in place of X^{q-1} in (3.109), we are then led to (3.114). $\square$

Remark 3.21 As X^{q-1} is symmetric, $\pi_k X^{q-1} = -B(X^{q-1})$ where $B(Y) = Y_- - Y_-^T$. This is the same minus sign that appears above in the comparison of Poisson brackets (see proof of Theorem 3.25; see also the discussion at the end of Section 1.5).

Remark 3.22 For $q = 2$, $H_2 = \frac{1}{2} \text{tr } X^2$, we recover the Toda lattice (1.1).

Remark 3.23 From Theorem 3.19, the Poisson-commutativity of the eigenvalues in the full matrix case follows from the same argument in the proof of the tridiagonal case, Theorem 3.17.

The introduction of variables $\{a_i, b_i\}$ in (3.3) by Flaschka and Manakov was simply an inspired guess. We now show that in fact the change of variables $\{x_i, y_i\} \mapsto \{a_i, b_i\}$ is a Poisson map. In particular, up to an irrelevant multiplicative constant, the physical symplectic structure in $\mathbb{R}^{2N}$ is essentially the Kirillov–Kostant symplectic form on the coadjoint orbit of Jacobi matrices of fixed given trace.

Remark 3.24 As $(\mathcal{O}_{X_0}, \omega_{X_0})$ is a symplectic manifold, any Hamiltonian H on $\mathcal{O}_{X_0}$ will generate a flow $X(t)$ that stays on the manifold. In particular,

3.4 The Toda Lattice: Full Matrix Case

if $X(0)$ is a Jacobi matrix on $\mathcal{O}_{X_0}$, then $X(t)$ will be a (tridiagonal) Jacobi matrix for all $t > 0$. A direct proof of this fact has already been given in Remark 3.2.

Theorem 3.25 *Let X_0 be an $N \times N$ Jacobi matrix. Then,*

$$\mathcal{O}_{X_0} = \{X : X \text{ is a Jacobi matrix such that } \operatorname{tr} X = \operatorname{tr} X_0\}. \quad (3.116)$$

Moreover, the Flaschka–Manakov map $\phi : \mathbb{R}^{2n} \to \{\text{Jacobi matrices}\}$ is Poisson, i.e., for smooth functions $F, G : \mathfrak{lo}^* \to \mathbb{R}$,

$$\{F \circ \phi, G \circ \phi\}_{\mathbb{R}^{2n}} = \{F, G\}_{X_0} \circ \phi. \quad (3.117)$$

Proof We leave the proof of (3.116) as an exercise for the reader (see Deift et al., 1986). To prove (3.117), let

$$\begin{cases} F(X) = a_i = \langle e_i, Xe_i \rangle, & 1 \le i \le N, \\ G(X) = X_{j,j} = \langle e_j, Xe_j \rangle, & 1 \le j \le N, \end{cases}$$

for $i \ne j$. Then, $\nabla F(X) = e_i e_i^T$ and $\nabla G(X) = e_j e_j^T$ and so $\pi_l \nabla F(X) = \nabla F(X)$, $\pi_l \nabla G(X) = \nabla G(X)$, from which we see that $[\pi_l \nabla F(X), \pi_l \nabla G(X)] = 0$. Hence,

$$\{a_i, a_j\}_{X_0}(X) = 0. \quad (3.118)$$

Now, let

$$\begin{cases} F(X) = b_i = \langle e_i, Xe_{i+1} \rangle, \\ G(X) = b_j = \langle e_j, Xe_{j+1} \rangle \end{cases}$$

for $i \ne j, 1 \le i, j \le N-1$. Then $\nabla F(X) = e_i e_{i+1}^T$ and $\nabla G(X) = e_j e_{j+1}^T$ and so $\pi_l \nabla F(X) = e_{i+1} e_i^T$ and $\pi_l \nabla G(X) = e_{j+1} e_j^T$. Hence,

$$[\pi_l \nabla F(X), \pi_l \nabla G(X)] = e_{i+1} e_i^T e_{j+1} e_j^T - e_{j+1} e_j^T e_{i+1} e_i^T$$
$$= \delta_{i,j+1} e_{i+1} e_j^T - \delta_{j,i+1} e_{j+1} e_i^T,$$

which implies

$$\{F, G\}_{X_0}(X) = \operatorname{tr}(X \delta_{i,j+1} e_{i+1} e_j^T - X \delta_{j,i+1} e_{j+1} e_i^T)$$
$$= \delta_{i,j+1} X_{j,i+1} - \delta_{j,i+1} X_{i,j+1}$$
$$= \delta_{i,j+1} X_{j,j+2} - \delta_{j,i+1} X_{i,i+2} = 0$$

for any Jacobi matrix X. Thus we have shown

$$\{b_i, b_j\}_{X_0}(X) = 0. \quad (3.119)$$

Finally, set
$$\begin{cases} F(X) = a_i, & 1 \leq i \leq N, \\ G(X) = X_{j,j+1}, & 1 \leq j \leq N-1. \end{cases}$$

Then $\nabla F(X) = e_i e_i^T$, $\nabla G(X) = e_j e_{j+1}^T$ and so $\pi_l \nabla F(X) = e_i e_i^T$, $\pi_l \nabla G(X) = e_{j+1} e_j^T$. Hence

$$[\pi_l \nabla F(X), \pi_l \nabla G(X)] = e_i e_i^T e_{j+1} e_j^T - e_{j+1} e_j^T e_i e_i^T$$
$$= \delta_{i,j+1} e_i e_j^T - \delta_{ji} e_{j+1} e_i^T,$$

and it follows that

$$\{F, G\}_{X_0}(X) = \text{tr}(X \delta_{i,j+1} e_i e_j^T - X \delta_{ji} e_{j+1} e_i^T)$$
$$= \delta_{i,j+1} X_{ji} - \delta_{ji} X_{i,j+1} = \delta_{i,j+1} X_{j,j+1} - \delta_{ji} b_i.$$

Thus

$$\{a_i, b_j\} = b_j(\delta_{i,j+1} - \delta_{ij}), \quad 1 \leq i \leq N, \quad 1 \leq j \leq N-1. \quad (3.120)$$

On the other hand, for $a_i = -y_i/2$, $b_i = \frac{1}{2} e^{\frac{1}{2}(x_i - x_{i+1})}$, we have, using $\{x_i, x_j\} = \{y_i, y_j\} = 0$, $\{x_i, y_j\} = \delta_{ij}$,

$$\{a_i, a_j\}_{\mathbb{R}^{2n}} = \left\{\frac{y_i}{2}, \frac{y_j}{2}\right\} = 0,$$
$$\{b_i, b_j\}_{\mathbb{R}^{2n}} = \left\{\frac{1}{2} e^{\frac{1}{2}(x_i - x_{i+1})}, \frac{1}{2} e^{\frac{1}{2}(x_j - x_{j+1})}\right\} = 0,$$

and

$$\{a_i, b_j\}_{\mathbb{R}^{2n}} = -\frac{1}{4}\left\{y_i, e^{\frac{1}{2}(x_j - x_{j+1})}\right\}$$
$$= -\frac{1}{4} e^{\frac{1}{2}(x_j - x_{j+1})}\left(-\frac{1}{2}\delta_{ij} + \frac{1}{2}\delta_{i,j+1}\right)$$
$$= \frac{1}{4} b_j(\delta_{ij} - \delta_{i,j+1}).$$

We see that these commutation relations agree with (3.118), (3.119), and (3.120), up to a trivial factor of -4, and (3.117) follows. □

3.5 Long-Time Behavior in the Full Hermitian Case

In order to analyze, in Chapters 5 and 6, the computation of the eigenvalues of complex matrix ensembles, in addition to real ensembles, we extend the Toda flows to Hermitian matrices.

3.5 Long-Time Behavior in the Full Hermitian Case

We now show that Moser's argument presented in Section 3.2 on the long time behavior of solutions of the Toda flow, applies in the full Hermitian matrix case. More precisely, in place of (1.1), $\partial_t X = [X, B(X)]$, $X(0) \in \Sigma_N$, we will consider more generally its complexification:

$$\partial_t X = [X, \widetilde{B}(X)], \quad X(0) = X_0 = X_0^* \in \Gamma_N, \qquad (3.121)$$

where Γ_N denotes the $N \times N$ Hermitian matrices and

$$\widetilde{B}(X) = X_- - X_-^* = -\widetilde{B}(X)^*.$$

Imitating the calculations of Section 1.2, one can show that (3.121) has a unique global solution, $X(t) = X(t)^*$ with $\mathrm{spec}(X(t)) = \mathrm{spec}(X(t)^*)$. Clearly, if $X_0 = \overline{X_0} = X_0^T$ then (3.121) reduces to (1.1).

Theorem 3.26 *Let $X(t)$, $t \geq 0$ solve 3.121 with $X(0) = X_0 = X_0^*$. Then, as $t \to \infty$, $X(t) \to \mathrm{diag}(\lambda_{\pi_1}, \ldots, \lambda_{\pi_N})$ where $\lambda_{\pi_1}, \ldots, \lambda_{\pi_N}$ is some permutation of the eigenvalues of X_0.*

Proof We have $\frac{dX}{dt} = X\widetilde{B}(X) + (X\widetilde{B}(X))^*$ and using the complex inner product for $\mathbb{C}^N$, we obtain

$$\partial_t X_{11} = 2\Re((e_1, X\widetilde{B}e_1)) = 2\Re((Xe_1, (X - X_{11})e_1)) = 2\sum_{i=1}^{N} |X_{i1}|^2 - 2|X_{11}|^2 \qquad (3.122)$$

$$= 2\sum_{i=2}^{N} |X_{i1}|^2. \qquad (3.123)$$

Following Moser's argument, we conclude that $X_{11}(t) \uparrow X_{11}(\infty) < \infty$, and

$$\sum_{i=2}^{N} |X_{i1}(t)|^2 \to 0.$$

Now,

$$\frac{d(X_{11} + X_{22})}{dt} = 2\Re((e_1, X\widetilde{B}e_1) + (e_2, X\widetilde{B}e_2))$$

$$= 2\Re(\sum_{i=2}^{N} |X_{i1}|^2 + \sum_{i=1}^{N} |X_{i2}|^2 - |X_{22}|^2 - 2|X_{12}|^2)$$

$$= 2\sum_{i=1}^{2} \sum_{2 < j \leq N} |X_{ji}(t)|^2,$$

and hence, as $t \to \infty$, $X_{11} + X_{22} \uparrow (X_{11} + X_{22})(\infty) < \infty$ and so

$$X_{22}(t) \to X_{22}(\infty) \quad \text{and} \quad \sum_{i=3}^{N} |X_{i2}(t)|^2 \to 0.$$

Continuing, we find, for $1 \le k \le N$,

$$\partial_t (X_{11} + \cdots + X_{k,k}) = 2 \sum_{i=1}^{k} \sum_{k<j \le N} |X_{ij}(t)|^2,$$

which leads to the desired result. □

Remark 3.27 Note that, unlike the Jacobi case, $X(t)$ is not always ordering, i.e., $\lambda_{\sigma(1)}, \ldots, \lambda_{\sigma(N)}$ is not always ordered. Indeed, if X_0 is diagonal, then $X(t) = X_0$ for all $t \ge 0$ and so no re-ordering can take place. We will discuss below a sufficient condition under which the flow with initial data in Γ_N is ordering.

Remark 3.28 We say an $n \times n$ matrix M is *complex Jacobi* if $M = M^*$, M is tridiagonal and $M_{i,i+1} \ne 0$ for $i = 1, \ldots, n-1$. The complexified flow (3.121) preserves complex Jacobi matrices, i.e., if $M(t = 0)$ is complex Jacobi, then so is the solution $M(t)$ of (3.121). Complex Jacobi matrices are related in a simple way to real Jacobi matrices. Indeed, if

$$M = \begin{pmatrix} a_1 & b_1 & & & \\ \overline{b_1} & a_2 & b_2 & & \\ & \overline{b_2} & a_3 & & \\ & & & \ddots & \\ & & & a_{N-1} & b_{N-1} \\ & & & \overline{b_{N-1}} & a_N \end{pmatrix}, \quad b_j = r_j e^{i\theta_j}, \ r_j > 0,$$

then

$$M = D^* J D,$$

where $D = \text{diag}(1, e^{i\phi_1}, \ldots, e^{i\phi_{N-1}})$, $\phi_j = \theta_1 + \cdots + \theta_j$, $j = 1, \ldots, N-1$ and

$$J = \begin{pmatrix} a_1 & r_1 & & & \\ r_1 & a_2 & r_2 & & \\ & r_2 & a_3 & & \\ & & & \ddots & \\ & & & a_{N-1} & r_{N-1} \\ & & & r_{N-1} & a_N \end{pmatrix}.$$

Moreover, it is easy to see that $M(t)$ solves the complexified flow (3.121) if and only if $J(t)$ solves the (real) Toda flow (1.1) and $\phi_i(t) = \phi_i(0), j = 1, \ldots, N-1$. In particular, the long time behavior of $M(t)$ can now be simply read off from the long time behavior of $J(t)$, as derived in Section 3.2.

Remark 3.29 The extended generalized flow, $p \geq 3$,

$$\partial_t X = [X, \tilde{B}(X^{p-1})], \quad X(0) = X_0 = X_0^*, \quad (3.124)$$

is not always diagonalizing. For example, consider the case $p = 3$,

$$X_0 = \begin{pmatrix} 0 & 1 \\ 1 & 0 \end{pmatrix} = X_0^* = X_0^T.$$

Then by (1.11), $X(t) = Q(t)^T X_0 Q(t)$ and $X(t)^{p-1} = X(t)^2 = I$, as $X_0^2 = I$. In particular, $\tilde{B}(X^{p-1}) = 0$ and so $X(t) = X_0$. However, it follows as in Chapter 2 that if $X(t)$ solves (3.124) for any $p \geq 2$, then

$$\partial_t X^{p-1} = [X^{p-1}, \tilde{B}(X^{p-1})],$$

which is the Toda flow. Hence $X^{p-1}(t)$ converges to the diagonal matrix

$$\mathrm{diag}(\lambda_{\sigma(1)}^{p-1}, \ldots, \lambda_{\sigma(N)}^{p-1}),$$

as $t \to \infty$. But, as above, $X(t)$ may not converge to a diagonal matrix.

Exercise 3.30 Does $X(t)$ always converge to some matrix under (3.124)?

3.6 The Generalized Toda Flow

It turns out that there is a remarkable way, discovered by Symes (1982), to solve (1.1) or (3.121) and the more general flows generated by the Hamiltonians such as $H_q(X) = -\frac{1}{q} \mathrm{tr} X^q$ explicitly. For more information, see for example, Deift et al. (1986); Symes (1980, 1982); Reyman and Semenov-Tian-Shansky (1979) and the references therein.

Exercise 3.31 (QR factorization) Let Y be an invertible complex matrix. Then Y has a unique factorization $Y = QR$, where Q is unitary and R is upper-triangular with $R_{ii} > 0, 1 \leq i \leq N$.

Theorem 3.32 *Suppose U is an open subset of $\mathbb{R}$ and X is a Hermitian matrix with $\sigma(X) \subset U$. Suppose that $g : U \to \mathbb{R}$ is continuous and define $g(X)$ by the spectral theorem, $g(X) = V^* g(D) V$, where $X = V^* D V$. Then the extended (and further) generalized flow*

$$\partial_t X = [X, \tilde{B}(g(X))], \quad X(0) = X_0 = X_0^* \in \Gamma_N, \quad (3.125)$$

has a global solution $X(t) = X^*(t)$, which can be computed as follows. Let

$$e^{tg(X_0)} = Q(t)R(t) \qquad (3.126)$$

be the QR factorization for $e^{tg(X_0)}$. Then

$$X(t) = Q^*(t) X_0 Q(t) \qquad (3.127)$$

and also

$$X(t) = R(t) X_0 R^{-1}(t). \qquad (3.128)$$

Moreover,

$$e^{tg(X(t))} = R(t) Q(t). \qquad (3.129)$$

Remark 3.33 The forms (3.128) and (3.129) for the solution of (3.125) has already appeared in Theorems 1.1 and 3.15 in special cases. There, the forms were obtained (implicitly) by integration. Here, the forms are obtained explicitly and algebraically by QR factorizations.

Proof As the QR factorization of a matrix Y is just the result of applying Gram–Schmidt to the columns of Y starting from the left, and $e^{tg(X_0)}$ is differentiable and nonsingular for all t, it follows that $Q(t)$ and $R(t)$ are differentiable functions of t. From (3.126) we have

$$g(X_0) Q R = \partial_t Q R + Q \partial_t R$$

or

$$g(\hat{X}(t)) \equiv Q^* g(X_0) Q = Q^* \partial_t Q + \partial_t R R^{-1}, \qquad (3.130)$$

where

$$\hat{X}(t) = Q^* X_0 Q. \qquad (3.131)$$

As $\partial_t R R^{-1}$ is upper triangular, it follows that

$$(Q^* \partial_t Q)_- = g(\hat{X}(t))_-$$

and, as Q is unitary, $Q^* \partial_t Q$ is skew-adjoint with a purely imaginary diagonal, i.e.,

$$Q^* \partial_t Q = (Q^* \partial_t Q)_- - (Q^* \partial_t Q)_-^* + D,$$

where D is diagonal and $D + D^* = 0$. From (3.130), as $g(\hat{X}(t))$ is Hermitian,

$$D = \text{diag}(g(\hat{X}(t))) - \text{diag}(\partial_t R R^{-1}),$$

is real and we must have $D = 0$. Thus we obtain

$$Q^* \partial_t Q = \tilde{B}(g(\hat{X}(t))),$$

3.6 The Generalized Toda Flow

or

$$\partial_t Q = Q \tilde{B}(g(\hat{X}(t))), \qquad Q(0) = I. \tag{3.132}$$

It follows that

$$\begin{aligned}
\partial_t \hat{X} &= \partial_t Q^* \, X_0 \, Q + Q^* \, X_0 \, \partial_t Q \\
&= -\tilde{B}\,(g(\hat{X}(t)))\, \hat{X} + \hat{X}\, \tilde{B}\,(g(\hat{X}(t))) \\
&= [\hat{X}, \tilde{B}(g(\hat{X}(t)))],
\end{aligned}$$

where $\hat{X}(0) = X_0$. As the solution of the differential equation is unique, $X(t) = \hat{X}(t)$. This completes the proof of (3.125). Formula (3.129) follows by conjugating (3.126) by $Q^*(t)$. $\square$

Corollary 3.34 *Suppose U and g are as in Theorem 3.32 and let $X(t)$ be the solution of (3.125). Then $e^{X(1)}$ can be computed by a QR step from $e^{X(0)}$:*

$$e^{g(X_0)} = Q(1)R(1) \to R(1)Q(1) = e^{g(X(1))}. \tag{3.133}$$

Proof In place of (3.126), consider the QR factorization

$$e^{sg(X(1))} = \tilde{Q}(s)R(s), \quad s \geq 0. \tag{3.134}$$

Then, as in Theorem 3.32,

$$\tilde{X}(s) = \tilde{Q}^*(s) X(1) \tilde{Q}(s)$$

solves (3.125), but with $\tilde{X}(0) = X(1)$. As in (3.129), we obtain

$$e^{g(X(1))} = \tilde{Q}(1)\tilde{R}(1) \to \tilde{R}(1)\tilde{Q}(1) = e^{g(\tilde{X}(1))}. \tag{3.135}$$

But $X(t)$ and $\tilde{X}(s)$ solve the same differential equation, and $X(1) = \tilde{X}(0)$, and so by uniqueness $X(1+s) = \tilde{X}(s)$. In particular, $\tilde{X}(1) = X(2)$ and so $e^{g(X(2))}$ is obtained by a QR step from $e^{g(X(1))}$, and so on. Iterating, $e^{g(X_k)}$ is obtained by performing k QR steps on $e^{g(X_0)}$. $\square$

In particular, setting $g(X) = X$ in Equation (3.125), we find, as noted by Symes (1982), the exponential of the solution of the Toda flow $e^{X(t)}$ at integer times $t = n$ can be obtained by a sequence of QR steps from e^{X_0}: also, setting $g(X) = \ln X$, we learn that the solution $X(t) = e^{\ln X(t)}$ of the QR flow (1.22), at integer times, can be obtained by a sequence of QR steps from $e^{\ln X_0} = X_0$. This is the Stroboscopic Property of Theorem 1.2.

We now prove the analog of (3.15), (3.16) for the extended generalized flow (3.125). This requires some interpretation as the eigenvalues of a general Hermitian matrix may not be simple and, in particular, the eigenvalues and eigenvectors may not be smooth functions of X. An appropriate choice, however, is given by the explicit solution of the flow given in the theorem above.

Theorem 3.35 *Let $X(t)$ solve the extended generalized Toda flow (3.124). Then there exists a smooth eigenbasis for $X(t)$ consisting of the columns of a unitary matrix $U^*(t)$ such that, for the first entries $U^*_{1j}(t)$ of the eigenvectors,*

$$U^*_{1j}(t) = \frac{U^*_{1j}(0)e^{\lambda_j^{q-1}t}}{\left(\sum_{i=1}^N |U^*_{1i}(0)|^2 e^{2\lambda_i^{q-2}t}\right)^{1/2}}. \qquad (3.136)$$

Moreover, for $q = 2$, in particular, we have

$$X_{11}(t) = \sum_{j=1}^N \lambda_j |U^*_{1j}(t)|^2, \qquad (3.137)$$

and

$$E(t) := \sum_{k=2}^N |X_{1k}(t)|^2 = \sum_{j=1}^N (\lambda_j - X_{11}(t))^2 |U^*_{1j}(t)|^2. \qquad (3.138)$$

Proof Choose an arbitrary spectral representation $X(0) = U^*(0)\Lambda U(0)$ and define the solution of (3.124) using Theorem 3.32,

$$X(t) = Q(t)^* X_0 Q(t) = Q(t)^* U^*(0) \Lambda\ U(0) Q(t),$$

where $Q(t)$ is obtained from the QR decomposition $e^{tX_0^{q-1}} = Q(t)R(t)$. Set $U^*(t) = Q(t)^* U^*(0)$: we consider its first row $U^*_1(t)$.

For any vector v, let $v^n = v/\|v\|$ denote its normalization. The first column of $Q(t)$ is the normalization of the first column of

$$e^{tX_0^{q-1}} = U^*(0)e^{t\Lambda^{q-1}}U(0):\qquad Q(t)e_1 = (U^*(0)e^{t\Lambda^{q-1}}U_1(0))^n.$$

Thus

$$U_1(t) = U(t)e_1 = U(0)Q(t)e_1 = U(0)\bigl(U^*(0)e^{t\Lambda^{q-1}}U_1(0)\bigr)^n = \bigl(e^{t\Lambda^{q-1}}U_1(0)\bigr)^n,$$

which is equation (3.136).

From the spectral representation $X(t) = U^*(t)\Lambda U(t)$, we have

$$X_{1k}(t) = \sum_{j=1}^N \lambda_j U^*_{1j}(t) U_{jk}(t),$$

3.6 The Generalized Toda Flow

and hence

$$E(t) = \sum_{k=2}^{N} |X_{1k}(t)|^2 = \sum_{k=2}^{N} X_{1k}(t) X_{k1}(t) = (X^2(t))_{11} - X_{11}^2(t)$$

$$= \sum_{k=1}^{N} \lambda_k^2 |U_{1k}^*(t)|^2 - \left(\sum_{k=1}^{N} \lambda_k |U_{1k}^*(t)|^2 \right)^2$$

$$= \sum_{k=1}^{N} (\lambda_k - X_{11}(t))^2 |U_{1k}^*(t)|^2,$$

which proves (3.138). □

As we will see, $E(t)$ provides error estimates on the top eigenvalue λ_1. Note

$$\lambda_1 - X_{11}(t) = \sum_{j=1}^{N} (\lambda_1 - \lambda_j) |U_{1j}^*(t)|^2. \tag{3.139}$$

Suppose the eigenvalues of X_0 are arranged in nonincreasing order

$$\lambda_1 = \lambda_2 = \cdots = \lambda_m > \lambda_{m+1} \geq \lambda_{m+2} \geq \cdots \geq \lambda_N. \tag{3.140}$$

Suppose $U_{1j}^*(0) \neq 0$ for some $1 \leq j \leq m$. Then without loss of generality, after permutation, we can assume that $U_{11}^*(t), \ldots, U_{1l}^*(t)$ are nonzero for some $1 \leq l \leq m$, and $U_{1,l+1}^*(t), \ldots, U_{1,m}^*(t) = 0$. It now follows from (3.136) that

$$U_{1j}^*(t) \to 0 \qquad \text{for } l+1 \leq j \leq N, \tag{3.141}$$

and

$$U_{1j}^*(t) \to \frac{U_{1j}^*(0)}{\left(\sum_{j=1}^{l} |U_{1j}^*(0)|^2 \right)^{1/2}} \qquad \text{for } 1 \leq j \leq l. \tag{3.142}$$

From (3.139), we see that

$$\lambda_1 - X_{11}(t) = \sum_{j=m+1}^{N} (\lambda_1 - \lambda_j) |U_{1j}^*(t)|^2 \tag{3.143}$$

and as $\lambda_j = \lambda_1$ for $1 \leq j \leq m$,

$$E(t) = \sum_{j=1}^{m} (\lambda_1 - X_{11}(t))^2 |U_{1j}^*(t)|^2 + \sum_{j=m+1}^{N} (\lambda_j - X_{11}(t))^2 |U_{1j}^*(t)|^2 \tag{3.144}$$

go to zero as $t \to \infty$.

While $\lambda_1 - X_{11}(t)$ is of course the true error in computing λ_1, we will use

$E(t) = \sum_{k=2}^{N} |X_{1k}(t)|^2$ to determine a convergence criterion as it is directly observable: Indeed, it follows from the min-max principle that if $E(t) < \varepsilon$ then $|X_{11}(t) - \lambda_i| < \varepsilon$ for some i. With high probability for a large class of random matrix ensembles, $\lambda_i = \lambda_1$ (see Proposition 6.5).

3.7 Convergence to the Top Eigenvalue

In this monograph we are interested in using the Toda algorithm to compute the top eigenvalue of a random real symmetric or Hermitian matrix. We do this by running the algorithm until $E(t) < \varepsilon$. As noted above this tells us that $|X_{11}(t) - \lambda_j| < \varepsilon$ for some eigenvalue λ_j of $X(t=0)$. So in order to conclude that λ_j is typically the top eigenvalue, we need to know that $X_{11}(t)$ converges to the top eigenvalue with high probability.

From the calculations leading up to (3.143), we see that the condition $U_{1j}^*(0) \neq 0$ for some $1 \leq j \leq m$, where

$$\lambda_1 = \lambda_2 = \cdots = \lambda_m > \lambda_{m+1} \geq \cdots \geq \lambda_N,$$

is sufficient for $X_{11}(t) \to \lambda_1$. Let $M \in \Gamma_N$ be an $N \times N$ Hermitian matrix with eigenvalues $\{\lambda_1 \geq \lambda_2 \geq \ldots \lambda_N\}$. Let

$$A_N = \{M \in \Gamma_N : \text{every eigenvector has nonzero first coordinate}\}.$$

The matrices in A_N must have simple spectrum. Indeed, two linearly independent eigenvectors u and v associated to an eigenvalue λ yield another eigenvector $v_1 u - u_1 v$ with first coordinate equal to zero. Clearly if $M \in A_N$ and $X(t)$ solves the extended Toda flow (3.121) with $X(0) = M$, then $X_{11}(t) \to \lambda_1$ as $t \to \infty$. Thus $M \in A_N$ is a sufficient condition for the Toda algorithm to compute the top eigenvalue of M.

Theorem 3.36 *The set A_N is a dense open set of full measure in Γ_N, i.e., the measure of $\Gamma_N \setminus A_N$ is 0.*

Proof Denote the $N \times N$ unitary matrices by U_N and by D_N the set of $N \times N$ diagonal matrices with real entries. From the spectral theorem, the map

$$F : U_N \times D_N \to \Gamma_N, \quad (U, D) \mapsto U^* D U$$

is surjective. Let $\tilde{U}_N \subset U_N$ consist of the matrices whose first column only have nonzero entries. The subset $\tilde{U}_N$ is of full measure, as the first column of a matrix in $\tilde{U}_N$ must avoid the coordinate planes of $\mathbb{C}^N$. Consider another subspace of full measure $\tilde{D}_N \subset D_N$, the set of real diagonal matrices with

3.7 Convergence to the Top Eigenvalue

simple spectrum. Again from the spectral theorem, $A_N = F(\tilde{U}_N \times \tilde{D}_N)$. Clearly, A_N is dense and open in Γ_N: We must show that its complement

$$F((U_N \setminus \tilde{U}_N) \times D_N) \cup F(U_N \times (D_N \setminus \tilde{D}_N))$$

is a set of measure zero. Both sets in the right-hand side are of measure zero, being the union of images of F of sets of measure zero. □

As noted earlier, the Toda algorithm is not necessarily an ordering algorithm on general full matrices in Σ_N or Γ_N. On the other hand, we have shown that if $X_0 \in A_N$ and $X(t)$ solves the Toda flow (3.121) with $X(0) = X_0$, then $X_{11}(t) \to \lambda_1$. We now describe a condition on X_0 that guarantees that $X_{22} \to \lambda_2$. To simplify notation, write $V(t) = U^*(t)$, so that $v_j = U^*(t)e_j$ is a smooth eigenvector for $X(t) = U^*(t)\Lambda U(t)$,

$$X(t)v_j(t) = \lambda_j v_j(t),$$

constructed as above to satisfy

$$\partial_t v_j(t) + \tilde{B}(X(t))v_j(t) = 0, \qquad (3.145)$$

(cf. equation 3.132). We conclude, as above (cf. 3.18), that

$$\partial_t v_{1j}(t) = (\lambda_j - X(t)_{11})v_{1j}(t).$$

Now

$$\begin{aligned}
\partial_t v_{2j}(t) &= -\langle e_2, \tilde{B}(X(t))v_j(t) \rangle \\
&= \langle \tilde{B}(X(t))e_2, v_j(t) \rangle \\
&= \langle X(t)e_2 - X_{22}(t)e_2 - 2\overline{X}_{21}e_1, v_j(t) \rangle \\
&= \langle (\lambda_j - X_{22}(t))e_2 - 2\overline{X}_{21}e_1, v_j(t) \rangle \\
&= (\lambda_j - X_{22}(t))v_{2j}(t) - 2X_{21}v_{1j}(t).
\end{aligned}$$

Thus for $j \neq k$

$$\begin{aligned}
\partial_t \left(v_{1j}(t)v_{2,k}(t) - v_{1k}(t)v_{2j}(t) \right) &= (\lambda_j - X_{11})v_{1j}v_{2,k} + v_{1j}\left((\lambda_k - X_{22})v_{2,k} - 2X_{21}v_{1k}\right) \\
&\quad - (\lambda_k - X_{11})v_{1k}v_{2j} - v_{1k}\left((\lambda_j - X_{22})v_{2j} - 2X_{21}v_{1j}\right) \\
&= (\lambda_j + \lambda_k - X_{11} - X_{22})(v_{1j}v_{2,k} - v_{1,k}v_{2j}).
\end{aligned}$$

Or, writing

$$v_{12,jk} := v_{1j}v_{2,k} - v_{1k}v_{2j}, \qquad (3.146)$$

we have

$$\partial_t v_{12,jk} = (\lambda_j + \lambda_k - X_{11} - X_{22})v_{12,jk}. \qquad (3.147)$$

Now observe that, using the orthonormality of $\{v_{1j}\}$ and $\{v_{2,k}\}$,

$$\sum_{1\leq i<l\leq N}(\lambda_i+\lambda_l)|v_{12,il}|^2$$

$$=\frac{1}{2}\sum_{1\leq i,l\leq N}(\lambda_i+\lambda_l)|v_{12,il}|^2$$

$$=\frac{1}{2}\sum_{1\leq i,l\leq N}(\lambda_i+\lambda_l)(v_{1,i}v_{2,l}-v_{1l}v_{2,i})\overline{(v_{1,i}v_{2,l}-v_{1l}v_{2,i})}$$

$$=\sum\lambda_i|v_{1,i}|^2+\sum\lambda_l|v_{2,l}|^2 \quad \text{(after cancellations)}$$

$$=X_{11}+X_{22}.$$

Thus we have for $1\leq j<k\leq N$,

$$\partial_t v_{12,jk}=\left(\lambda_j+\lambda_k-\left(\sum_{1\leq i<l\leq N}(\lambda_i+\lambda_l)|v_{12,il}|^2\right)\right)v_{12,jk}. \tag{3.148}$$

Using the fact that

$$\sum_{1\leq j<k\leq N}v_{12,jk}^2=\frac{1}{2}\sum_{1\leq j<k\leq N}(v_{1j}v_{2k}-v_{1k}v_{2j})^2=1,$$

the same argument that produced (3.16) using (3.20) but now using (3.148), yields the following.

Theorem 3.37 *Let $X(t)$ solve the extended Toda flow (3.124). Let*

$$v_{12,jk}=v_{1j}v_{2,k}-v_{1k}v_{2j}. \tag{3.149}$$

Then

$$v_{12,jk}(t)=\frac{v_{12,jk}(0)e^{t(\lambda_j+\lambda_k)}}{\left(\sum_{1\leq i<l\leq M}|v_{12,il}|^2 e^{2t(\lambda_i+\lambda_l)}\right)^{1/2}}. \tag{3.150}$$

What is going on here? Consider the space of skew 2-tensors $\Lambda_2=\Lambda_2(\mathbb{C}^N)$, spanned by the lexicographically ordered basis

$$e_1\wedge e_2,\ e_1\wedge e_3,\ \ldots,\ e_1\wedge e_N,\ e_2\wedge e_3,\ \ldots \tag{3.151}$$

of dimension $N(N-1)/2$. We write $(i,j)<(k,\mathfrak{l}\mathfrak{o})$ if $e_i\wedge e_j$ appears before $e_k\wedge e_\ell$ in the lexicographic order, i.e., $i<k$ or $i=k, j<\ell$. We write $(k,\ell)>(i,j)$ if $(i,j)<(k,\ell)$.

The space Λ_2 carries a natural inner product,

$$\langle u\wedge v,\tilde{u}\wedge\tilde{v}\rangle=\det\begin{pmatrix}\langle u,\tilde{u}\rangle & \langle u,\tilde{v}\rangle \\ \langle v,\tilde{u}\rangle & \langle v,\tilde{v}\rangle\end{pmatrix},$$

3.7 Convergence to the Top Eigenvalue

where the inner product in $\mathbb{C}^N$ is $\langle u, v \rangle = \sum_j \bar{u}_j v_j$.

Proposition 3.38 *The operator $T_2 = T \otimes I + I \otimes T$ acts in Λ_2,*

$$T_2 u \wedge v = (Tu) \wedge v + u \wedge (Tv).$$

If $T = T^*$, then T_2 is self-adjoint in Λ_2. If $\{v_j\}$ is an orthonormal basis of eigenvectors for T, $Tv_j = \lambda_j v_j$, then $\{v_i \wedge v_j, i < j\}$ is a complete orthonormal basis of eigenvectors for T_2 in Λ_2 with associated eigenvalues $\lambda_i + \lambda_j$, respectively.

Proof The proof that $T_2 : \Lambda_2 \to \Lambda_2$ is well defined is left to the reader. We verify the eigenvalue–eigenvector relations:

$$T_2(v_i \wedge v_j) = (Tv_i) \wedge v_j + v_i \wedge (Tv_j)$$
$$= \lambda_i(v_i \wedge v_j) + \lambda_j(v_i \wedge v_j) = (\lambda_i + \lambda_j) v_i \wedge v_j. \quad \square$$

Commensurate with the lexicographic ordering in (3.151), entries of vectors in Λ_2 are labeled by pairs $(i, j), i < j$. In a similar fashion, matrix entries of a linear map $X_2 : \Lambda_2 \to \Lambda_2$ are labeled by couples of pairs $((i, j), (k, \ell); i < j, k < \ell)$. From the ordering of the basis, we define in an obvious way diagonal and triangular matrices: lower triangular, diagonal, and upper triangular entries correspond, respectively, to

$$e_i \wedge e_j > e_k \wedge e_\ell, \quad e_i \wedge e_j = e_k \wedge e_\ell, \quad e_i \wedge e_j < e_k \wedge e_\ell.$$

An operation B_2 takes a linear map $X_2 : \Lambda_2 \to \Lambda_2$ to a skew Hermitian matrix $B_2(X_2)$, with the same lower triangular entries of X_2 and diagonal consisting of the imaginary entries of the diagonal entries of X_2.

Proposition 3.39 *For $T \in \Gamma_N$,*

$$B_2(T \otimes I + I \otimes T) = B(T) \otimes I + I \otimes B(T).$$

Proof We compare matrix entries $((i, j), (k, \ell))$ of both sides,

$$\text{LHS} = \langle e_i \wedge e_j, B_2(T \otimes I + I \otimes T) e_k \wedge e_\ell \rangle,$$
$$\text{RHS} = \langle e_i \wedge e_j, (B(T) \otimes I + I \otimes B(T)) e_k \wedge e_\ell \rangle$$
$$= \begin{pmatrix} \langle e_i, B(T)e_k \rangle & \langle e_i, e_\ell \rangle \\ \langle e_j, B(T)e_k \rangle & \langle e_j, e_\ell \rangle \end{pmatrix} + \begin{pmatrix} \langle e_i, e_k \rangle & \langle e_i, B(T)e_\ell \rangle \\ \langle e_j, e_k \rangle & \langle e_j, B(T)e_\ell \rangle \end{pmatrix}$$
$$= \delta_{j\ell}(B(T))_{ik} - \delta_{i\ell}(B(T))_{jk} + \delta_{ik}(B(T))_{j\ell} - \delta_{jk}(B(T))_{i\ell}.$$

Case 1: Lower triangular entries, $i > k$ or $i = k, j > \ell$.

Lower triangular entries of $B_2(X_2)$ and X_2 are equal, and thus

$$\text{LHS} = \langle e_i \wedge e_j, (T \otimes I + I \otimes T) e_k \wedge e_\ell \rangle = \delta_{j\ell} T_{ik} - \delta_{i\ell} T_{jk} + \delta_{ik} T_{j\ell} - \delta_{jk} T_{i\ell}.$$

Suppose first that $i > k$. Then LHS $= \delta_{j\ell} T_{ik} - \delta_{il} T_{jk}$. Also, as lower triangular entries of $B(T)$ and T are equal,

$$\text{RHS} = \delta_{jl}(B(T))_{ik} - \delta_{i\ell}(B(T))_{jk} = \delta_{j\ell} T_{ik} - \delta_{i\ell} T_{jk},$$

and LHS = RHS. If $i = k, j > \ell$, we have $i > \ell$: again,

$$\text{LHS} = T_{j\ell}, \quad \text{RHS} = (B(T))_{j\ell} = T_{j\ell}.$$

Case 2: Upper triangular entries, $i < k$ or $i = k, j < \ell$.

Upper triangular entries of the pairs $B_2(X_2)$ and X_2 and $B(T)$ and T now have opposite signs.

$$\text{LHS} = -(\delta_{j\ell} T_{ik} - \delta_{i\ell} T_{jk} + \delta_{ik} T_{j\ell} - \delta_{jk} T_{i\ell}).$$

Suppose first $i < k$. Then LHS $= -\delta_{j\ell} T_{ik} + \delta_{jk} T_{i\ell}$, and equality holds again,

$$\text{RHS} = \delta_{j\ell}(B(T))_{ik} - \delta_{jk}(B(T))_{i\ell} = -\delta_{j\ell} T_{ik} + \delta_{jk} T_{i\ell}.$$

If $i = k, j < \ell$, we are also fine:

$$\text{LHS} = -T_{j\ell};$$
$$\text{RHS} = (B(T))_{j\ell} = -T_{j\ell}.$$

Case 3: Diagonal entries, $i = k$ and $j = \ell$, have a special treatment:

$$\text{LHS} = \langle e_i \wedge e_j, B_2(T \otimes I + I \otimes T) e_i \wedge e_j \rangle$$
$$= \langle e_i \wedge e_j, (T \otimes I + I \otimes T) e_i \wedge e_j \rangle$$
$$= \Im(T_{ii} + T_{jj}), \text{RHS} = (B(T))_{ik} + (B(T))_{j\ell} = \Im T_{ii} + \Im T_{jj},$$

and the proof is complete. □

Theorem 3.40 *If $X(t)$ solves $\partial_t X(t) = [X(t), B(X(t))]$, then $X_2(t) = X(t) \otimes I + I \otimes X(t)$ satisfies*

$$\partial_t X_2(t) = [X_2(t), B_2(X_2(t))], \quad \text{for } B_2(X_2(t)) = B(X(t)) \otimes I + I \otimes B(X(t)).$$

Proof We have

$$\partial_t X_2(t) = (\partial_t X \otimes I + I \otimes \partial_t X)(u \wedge v) = \partial_t X u \wedge v + u \wedge \partial_t X v \quad (3.152)$$

$$= XBu \wedge v - BXu \wedge v + u \wedge XBv - u \wedge BXu,$$

3.7 Convergence to the Top Eigenvalue

and, using the preceding proposition,

$[X_2, B_2(X_2)](u \wedge v) = X_2 (Bu \wedge v + u \wedge Bv) - B_2(X_2)(Xu \wedge v + u \wedge Xv) =$

$XBu \wedge v + Bu \wedge Xv + Xu \wedge Bv + u \wedge XBu - BXu \wedge v - Xu \wedge Bv - Bu \wedge Xv - u \wedge BXu,$

which is the right-hand side of equation (3.152). □

Now the first component of $v_i \wedge v_j$ in the ordered basis $(e_i \wedge e_j)_{i<j}$ is

$$\langle e_1 \wedge e_2, v_i \wedge v_j \rangle = \det \begin{pmatrix} \langle e_1, v_i \rangle & \langle e_1, v_j \rangle \\ \langle e_2, v_i \rangle & \langle e_2, v_j \rangle \end{pmatrix}$$

$$= v_{1,i} v_{2,j} - v_{1,j} v_{2,i},$$

which is exactly $v_{12,ij}$. Also (cf. (3.137)),

$$\langle e_1 \wedge e_2, X_2 e_1 \wedge e_2 \rangle = \sum_{1 \leq i < l \leq N} (\lambda_i + \lambda_l) |v_{12,il}|^2. \tag{3.153}$$

Thus (3.148) can be written in the form

$$\partial_t v_{12,jk} = \left((\lambda_j + \lambda_k) - \langle e_1 \wedge e_2, X_2(e_1 \wedge e_2) \rangle \right) v_{12,jk}. \tag{3.154}$$

Theorem 3.40 explains (3.154), i.e., if $X(t)$ solves the Toda equation $\partial_t X(t) = [X, B(X)]$ in Γ_N, then $X_2 = X \otimes I + I \otimes X$ solves the Toda equation in Λ_2,

$$\partial_t X_2 = [X_2, B_2(X_2)],$$

and (3.154) is just the evolution equation for the first components of the eigenvectors of X_2 (cf. (3.20)). We have also proved the following:

Theorem 3.41 *Let $A_N^{(2)}$ be the set of $X \in \Gamma_N$ such that if u is an eigenvector of X for some eigenvalue λ, then $u_1 \neq 0$, and if $u \wedge v$ is an eigenvector of $X_2 = X \otimes I + I \otimes X$ for some eigenvalue $\lambda + \mu$, then $\langle e_1 \wedge e_2, u \wedge v \rangle \neq 0$.*
Suppose $X(t)$ solves Toda with $X(0) = X_0 \in A_N^{(2)}$. Then as $t \to \infty$,

$$X_{11}(t) \to \lambda_1, \qquad X_{22}(t) \to \lambda_2.$$

Proof For $X \in A_N^{(2)} \subset A_N$, we necessarily have $\lambda_1 > \lambda_2 > \cdots > \lambda_N$. We have already shown, using (3.141) et seq. that $X_{11}(t) \to \lambda_1$. But from (3.150), we have

$$v_{12,jk}(t) = \frac{v_{12,jk}(0) e^{t(\lambda_j + \lambda_k)}}{\left(\sum_{1 \leq i < l \leq M} |v_{12,il}(0)|^2 e^{2t(\lambda_i + \lambda_l)} \right)^{1/2}}, \qquad t \geq 0. \tag{3.155}$$

From this it follows, again as in the $\mathbb{C}^N$ case, that

$$\langle e_1 \wedge e_2, \Lambda_2(X_2(t))(e_1 \wedge e_2) \rangle \to \lambda_1 + \lambda_2,$$

as $\lambda_1 + \lambda_2 > \lambda_i + \lambda_j$ for all $(i,j) \neq (1,2)$ and $(e_1 \wedge e_2, v_1 \wedge v_2) = v_{12,12} \neq 0$. Here we have used (3.152) and the assumption that $X_2 \in A_N^{(2)}$. Thus

$$X_{11}(t) + X_{22}(t) = \langle e_1 \wedge e_2, X_2(t)(e_1 \wedge e_2)\rangle \to \lambda_1 + \lambda_2,$$

from which we conclude that $X_{22}(t) \to \lambda_2$. □

Exercise 3.42 Let $X \in \Gamma_N$ with eigenvalues $\lambda_1 > \lambda_2 > \cdots > \lambda_N$ and normalized eigenvectors $v^{(j)}$, $Xv^{(j)} = \lambda_j v^{(j)}$. For $1 \le j_1 < j_2 < \cdots < j_k \le N$, set

$$d_{(j_1,\ldots,j_k)}(t) = \det\begin{pmatrix} v_{1j_1} & v_{2j_1} & \cdots & v_{kj_1} \\ v_{1j_2} & v_{2j_2} & \cdots & v_{kj_2} \\ \vdots & & & \vdots \\ v_{1j_k} & v_{2j_k} & \cdots & v_{kj_k} \end{pmatrix}.$$

Show that if $X = X(t)$ solves Toda then

$$d_{(j_1,\ldots,j_k)}(t) = \frac{d_{(j_1,\ldots,j_k)}(0) \, e^{t(\lambda_{j_1}+\cdots+\lambda_{j_k})}}{\left(\sum_{1\le i_1<i_2<\cdots<i_k\le N} d^2_{(i_1,\ldots,i_k)}(0) \, e^{2t(\lambda_{i_1}+\cdots+\lambda_{i_k})}\right)^{1/2}},$$

generalizing equation (3.155).

Exercise 3.43 Show that $A_N^{(2)}$ is an open dense set of full measure in $\mathbb{R}^{N^2}$.

Exercise 3.44 Generalize Theorem 3.41, to give a condition for the Toda flow to be ordering on an open dense set of full measure in $\mathbb{R}^{N^2}$.

3.8 Action-Angle Variables for the Toda Flow on Jacobi Matrices

We return to the Toda lattice on Jacobi matrices. We have shown that it is a completely integrable system in the sense of Liouville. We now construct the *actions* and the *angles* for the flow. Let $X_0 \in \mathcal{J}_N$, the set of Jacobi matrices. We have shown that the eigenvalues $\lambda_1 > \cdots > \lambda_N$ of X_0 are integrals of the motion $X(t)$ with initial condition $X(0) = X_0$. They are independent and Poisson-commute (see Theorems 3.18 and 3.17).

Let $\mu_k = \ln u_{1k}(t)$ where u_{1k} is, as above, the first component of the eigenvector $u_k(t)$ of $X(t)$ corresponding to the eigenvalue λ_k. We compute the Poisson bracket $\{\mu_k, \lambda_j\}$, $1 \le k, j \le N$. From Equation (3.132) it follows that under the generalized flow generated by the Hamiltonian $H_p(X) =$

3.8 Action-Angle Variables for Toda on Jacobi Matrices

$\frac{1}{p} \operatorname{tr} X^p$,

$$\{\mu_k, H_p\} = \frac{1}{u_{1k}(t)} \partial_t u_{1k} = \lambda_k^{p-1} - (X(t)^{p-1})_{11}. \tag{3.156}$$

Set

$$\nu_k = \ln(u_{1k}(t)/u_{1N}(t)) = \mu_k - \mu_N, \quad 1 \le k \le N-1. \tag{3.157}$$

Then

$$\{\nu_k, H_p\} = \lambda_k^{p-1} - \lambda_N^{p-1}, \quad 1 \le k \le N-1, \tag{3.158}$$

or using $H_p = \frac{1}{p} \sum_{i=1}^N \lambda_i^p$,

$$\sum_{i=1}^N \lambda_i^{p-1} \{\nu_k, \lambda_i\} = \lambda_k^{p-1} - \lambda_N^{p-1}. \tag{3.159}$$

Let $V = V(\lambda)$ be the Vandermonde matrix defined by

$$V(\lambda) = \begin{pmatrix} 1 & 1 & \cdots & 1 \\ \lambda_1 & \lambda_2 & \cdots & \lambda_N \\ \vdots & \vdots & & \vdots \\ \lambda_1^{N-1} & \lambda_2^{N-1} & \cdots & \lambda_N^{N-1} \end{pmatrix},$$

so that $V^{-1}(1, \lambda_i, \ldots, \lambda_i^{N-1})^* = e_i$, $1 \le i \le N$. For fixed k, let

$$\pi_i = \{\nu_k, \lambda_i\}, 1 \le i \le N.$$

Then, with $\pi = (\pi_1, \ldots, \pi_N)^*$, (3.158) takes the form

$$V\pi = (\lambda_k^0 - \lambda_N^0, \ldots, \lambda_k^{N-1} - \lambda_N^{N-1})^*,$$

and so

$$\pi = V^{-1} \begin{pmatrix} 1 \\ \lambda_k \\ \vdots \\ \lambda_k^{N-1} \end{pmatrix} - V^{-1} \begin{pmatrix} 1 \\ \lambda_N \\ \vdots \\ \lambda_N^{N-1} \end{pmatrix} = e_k - e_N, \quad 1 \le k \le N-1.$$

It follows that

$$\{\nu_k, \lambda_i\} = \delta_{ik} - \delta_{iN}, \quad 1 \le k \le N-1, \ 1 \le i \le N. \tag{3.160}$$

In particular

$$\{\nu_k, \lambda_i\} = \delta_{ik}, \quad 1 \le i, k \le N-1. \tag{3.161}$$

What do we know about the Poisson brackets $\{v_k, v_l\}$? From the Jacobi identity for $1 \le k, l \le N-1$ and $1 \le i \le N$,

$$\{\{v_k, v_l\}, \lambda_i\} + \{\{v_l, \lambda_i\}, v_k\} + \{\{\lambda_i, v_k\}, v_l\} = 0. \tag{3.162}$$

But $\{v_l, \lambda_i\}$ is a constant and hence $\{\{v_l, \lambda_i\}, v_k\} = 0$ for all $1 \le i \le N$. Similarly, $\{\{\lambda_i, v_k\}, v_l\} = 0$. It follows from (3.162) that

$$\{\{v_k, v_l\}, \lambda_i\} = 0. \tag{3.163}$$

Exercise 3.45 Prove: A symplectic manifold of dimension $2N$ cannot have a set of p independent Poisson-commuting functions with $p > N$.

It follows from (3.163) and Theorems 3.17 and 3.18 that $\{v_k, v_l\}$ must be functions of $\lambda_1, \ldots, \lambda_N$.

$$\{v_k, v_l\} = F_{kl}(\lambda_1, \ldots, \lambda_N).$$

We would like to conclude that this is 0, but unfortunately it is not. The appropriate commuting variables turn out to be

$$\theta_k := v_k + \frac{1}{2}\ln\left|\frac{P'(\lambda_k)}{P'(\lambda_N)}\right| = \ln\left(\frac{u_{1k}}{u_{1N}}\left|\frac{P'(\lambda_k)}{P'(\lambda_N)}\right|^{1/2}\right), \quad 1 \le k \le N-1, \tag{3.164}$$

where

$$P(\lambda) := \det(\lambda - X). \tag{3.165}$$

Note that as the spectrum of Jacobi matrix is simple, $P'(\lambda_k) \ne 0, 1 \le k \le N$. Note further that as $\ln\left|\frac{P'(\lambda_k)}{P'(\lambda_N)}\right|$ involves only eigenvalues, we still have

$$\{\theta_k, \lambda_i\} = \delta_{ik}, \quad 1 \le i, k \le N-1, \tag{3.166}$$

but now, as we will eventually show,

$$\{\theta_k, \theta_j\} = 0, \quad 1 \le k, j \le N-1. \tag{3.167}$$

The variables θ_k were introduced in Deift et al. (1986) and give commuting variables as in (3.167), not only for the Toda flow on tridiagonal matrices but also in the case of the flow on full $N \times N$ symmetric matrices. The proof of (3.167) in Deift et al. (1986) is lengthy and involves group theoretic properties of the Toda flow on full $N \times N$ matrices. Rather than reproducing this proof, we give a direct proof of (3.167) for the θ_k, which works for the Jacobi case at hand (the proof begins with (3.186)).

3.8 Action-Angle Variables for Toda on Jacobi Matrices

The variables θ_k have a determinantal representation. By Cramer's rule,

$$\left\langle e_1, \frac{1}{\lambda - X} e_1 \right\rangle = \frac{D(X, \lambda)}{P(\lambda)}, \tag{3.168}$$

where $P(\lambda) = \det(\lambda - X)$ and $D(X, \lambda)$ is the determinant of the $(N-1) \times (N-1)$ matrix obtained by deleting the first row and column of $\lambda - X$. On the other hand, by the spectral theorem,

$$\left\langle e_1, \frac{1}{\lambda - X} e_1 \right\rangle = \sum_{k=1}^{N} \frac{u_{1k}^2}{\lambda - \lambda_k}. \tag{3.169}$$

Equating (3.168) and (3.169), and letting $\lambda \to \lambda_k$, we obtain

$$\frac{D(X, \lambda_k)}{P'(\lambda_k)} = u_{1k}^2,$$

or

$$D(X, \lambda_k) = u_{1k}^2 P'(\lambda_k), \quad k = 1, \ldots, N, \tag{3.170}$$

so

$$\theta_k = \frac{1}{2} \ln \left| \frac{D(X, \lambda_k)}{D(X, \lambda_N)} \right|, \quad k = 1, \ldots, N-1. \tag{3.171}$$

Note that

$$\{v_k, \sum_{i=1}^{N} \lambda_i\} = \sum_{i=1}^{N} \{v_k, \lambda_i\} = \sum_{i=1}^{N} (\delta_{ik} - \delta_{iN}) = 0, \tag{3.172}$$

and so, in addition to (3.166), we also have

$$\{\theta_k, \lambda_i - \frac{1}{N} \sum_{l=1}^{N} \lambda_l\} = \delta_{ik}, \quad 1 \le i, k \le N-1. \tag{3.173}$$

Thus, we have $N-1$ angles $\{\theta_k\}_{k=1}^{N-1}$ and $N-1$ actions $\{\lambda_k\}_{k=1}^{N-1}$. In order to prove the integrability of the original Toda Hamiltonian in $\mathbb{R}^{2n}$, we need another angle and another action so that $2(N-1) + 2 = 2N = \dim \mathbb{R}^{2N}$. These are provided by the center of mass for the system

$$x_{\text{cm}} := \frac{1}{N}(x_1 + \cdots + x_N), \tag{3.174}$$

and the total momentum

$$y_T := y_1 + \cdots + y_N. \tag{3.175}$$

As $y_T = -2(a_1 + \cdots + a_N) = -2\operatorname{tr} X = -2\sum_{i=1}^{N} \lambda_i$, y_T is clearly conserved by the Toda flow. We have

$$\{x_{cm}, y_T\} = \frac{1}{N}\sum_i\sum_j\{x_i, y_j\} = 1.$$

Also for any j, using Exercise 3.7,

$$\{x_{cm}, \lambda_j\} = \frac{1}{N}\sum_{i=1}^{N}\partial_{y_i}\lambda_j = -\frac{1}{2N}\sum_{i=1}^{N}\partial_{a_i}\lambda_j = -\frac{1}{2N}\sum_{i=1}^{N}u_j^2(i) = -\frac{1}{2N},$$

i.e.,

$$\{x_{cm}, \lambda_j\} = -\frac{1}{2N}. \qquad (3.176)$$

Thus,

$$\{x_{cm}, \lambda_j - \frac{1}{N}\sum\lambda_l\} = -\frac{1}{2N} + \frac{1}{2N}\{x_{cm}, y_T\} = 0. \qquad (3.177)$$

And as $y_T = -2\sum_{i=1}^{N}\lambda_i$, we immediately have

$$\{y_T, \lambda_j - \frac{1}{N}\sum_{l=1}^{N}\lambda_l\} = 0. \qquad (3.178)$$

Also, from (3.172),

$$\{v_k, \sum_{i=1}^{N}\lambda_i\} = 0, \quad 1 \le k \le N-1,$$

and so

$$\{v_k, y_T\} = 0. \qquad (3.179)$$

Finally, we have

$$\{v_k, x_{cm}\} = 0. \qquad (3.180)$$

This is because v_k is built purely out of the entries of X, which do not depend on x_{cm}, but only on the differences $x_i - x_{i+1}$, $1 \le i \le N-1$. Indeed, for the flow generated by the Hamiltonian $K = x_{cm}$, we have

$$\partial_t x_i = \{x_i, x_{cm}\} = 0,$$

$$\partial_t y_i = \{y_i, x_{cm}\} = \frac{1}{N}\{y_i, \sum_{l=1}^{N} x_l\} = -\frac{1}{N},$$

and so

$$x_i(t) = x_i(0),$$

$$y_i(t) = y_i(0) - \frac{t}{N}, \quad \text{i.e., } a_i = a_i(0) + \frac{t}{2N},$$

3.8 Action-Angle Variables for Toda on Jacobi Matrices

which implies
$$X(t) = X(0) + \frac{t}{2N}I. \tag{3.181}$$

Thus the eigenvectors of $X(t)$ do not change, but all the eigenvalues of $X(t)$ are shifted by the same amount. Hence
$$\{v_k, x_{\text{cm}}\} = \partial_t v_k = 0,$$

and as $P'(\lambda_k) = \prod_{q \neq k}(\lambda_k - \lambda_q)$ only depends on the differences of the eigenvalues, it follows from (3.164) that
$$\{\theta_k, x_{\text{cm}}\} = 0, \quad 1 \le k \le N-1.$$

We have the following result. For $1 \le i \le N-1$, set $\tilde{\lambda}_i = \lambda_i - \frac{1}{N}\sum_{j=1}^{N}\lambda_j$, and also set $\tilde{\lambda}_0 = y_T$, $\theta_0 = x_{\text{cm}}$.

Theorem 3.46 *The map* $\Phi : \mathbb{R}^{2N} \to \mathbb{R}^{2N}$ *taking*
$$(x, y) \mapsto (\theta_0, \theta_1, \dots, \theta_{N-1}, \tilde{\lambda}_0, \tilde{\lambda}_1, \dots, \tilde{\lambda}_{N-1})$$

is a symplectomorphism, i.e., a symplectic diffeomorphism. Moreover, the variables on the right-hand side are action-angle variables for the Toda flow, i.e.,
$$\begin{cases} \{\theta_i, \theta_j\} = 0, & 0 \le i, j \le N-1, \\ \{\tilde{\lambda}_i, \tilde{\lambda}_j\} = 0, & 0 \le i, j \le N-1, \\ \{\theta_i, \tilde{\lambda}_j\} = \delta_{ij}, & 0 \le i, j \le N-1, \end{cases} \tag{3.182}$$

and under the Toda flow,
$$x_{\text{cm}} = \theta_0(t) = x_{\text{cm}}(0) + \frac{1}{N}y_T(0)t, \tag{3.183}$$

$$\theta_i(t) = \theta_i(0) + 4\left(\tilde{\lambda}_i + \sum_{j=1}^{N-1}\tilde{\lambda}_j\right)t, \quad 1 \le i \le N-1, \tag{3.184}$$

and
$$\tilde{\lambda}_i(t) = \tilde{\lambda}_i(0), \quad 0 \le i \le N-1. \tag{3.185}$$

Proof The relations (3.182) have already been proved. As $H_{\text{Toda}} = 2\operatorname{tr} X^2$, we have from (3.176),
$$\partial_t \theta_0 = \{x_{\text{cm}}, 2\sum_{i=1}^{N}\lambda_i^2\} = 4\sum_{i=1}^{N}\lambda_i\{x_{\text{cm}}, \lambda_i\}$$
$$= -\frac{2}{N}\sum_{i=1}^{N}\lambda_i = \frac{1}{N}y_T(0),$$

and from (3.160), for $1 \le i \le N-1$,

$$\partial_t \theta_i = \{\theta_i, 2\sum_{k=1}^N \lambda_k^2\} = 4\sum_{k=1}^N \lambda_k \{\theta_i, \lambda_k\}$$

$$= 4\sum_{k=1}^N \lambda_k(\delta_{ik} - \delta_{kN}) = 4(\lambda_i - \lambda_N).$$

A simple algebraic calculation now shows that $(\lambda_i - \lambda_N) = \tilde{\lambda}_i + \sum_{j=1}^{N-1} \tilde{\lambda}_j$. Integrating we obtain (3.183), (3.184), and (3.185) is immediate. This shows that Φ linearizes the Toda flow.

We now show that Φ is a diffeomorphism. Suppose $\Phi(x, y) = \Phi(x', y')$ for some $(x, y), (x', y') \in \mathbb{R}^{2N}$. Then

$$\sum_{i=1}^N \lambda_i = -\frac{1}{2}y_T = -\frac{1}{2}y'_T = \sum_{i=1}^N \lambda'_i,$$

and

$$\tilde{\lambda}_i = \lambda_i - \frac{1}{N}\sum_{j=1}^N \lambda_j = \lambda'_i - \frac{1}{N}\sum_{j=1}^N \lambda'_j, \qquad 1 \le i \le N-1.$$

But as $\sum \lambda_j = \sum \lambda'_j$, we see that $\lambda_i = \lambda'_i$ for $1 \le i \le N-1$, and using $\sum \lambda_j = \sum \lambda'_j$ again, we conclude $\lambda_N = \lambda'_N$. Furthermore, $\theta_i = \theta'_i$ implies

$$\ln\left(\frac{u_{1,i}}{u_{1N}}\left|\frac{P'(\lambda_i)}{P'(\lambda_N)}\right|^{1/2}\right) = \ln\left(\frac{u'_{1,i}}{u'_{1N}}\left|\frac{P'(\lambda'_i)}{P'(\lambda'_N)}\right|^{1/2}\right),$$

and as $\lambda_i = \lambda'_i, 1 \le i \le N$, we have

$$\frac{u_{1,i}}{u_{1N}} = \frac{u'_{1,i}}{u'_{1N}}, \qquad 1 \le i \le N-1,$$

and so

$$\frac{1}{u_{1N}^2} = \sum_{i=1}^{N-1}\frac{u_{1,i}^2}{u_{1N}^2} + 1 = \sum_{i=1}^{N-1}\frac{u'^2_{1,i}}{u'^2_{1N}} + 1 = \frac{1}{u'^2_{1N}},$$

and hence $u_{1N} = u'_{1N}$. But then $u_{1,i} = u'_{1,i}$ for $1 \le i \le N-1$.

It follows then from Lemma 3.8 that the Flaschka matrices X and X' associated with (x, y) and (x', y'), respectively, are equal. Thus $y_i = -2a_i = -2a'_i = y'_i$ and $b_i = b'_i$ implies $x_i - x_{i+1} = x'_i - x'_{i+1}$, which implies in turn that $x_i - x_1 = x'_i - x'_1$ for $1 \le i \le N$. Thus,

$$\sum_{i=1}^N x_i - Nx_1 = \sum_{i=1}^N x'_i - Nx'_1.$$

3.8 Action-Angle Variables for Toda on Jacobi Matrices   93

But $x_{cm} = \theta_0 = \theta'_0 = x'_{cm}$ and so $\sum_{i=1}^{N} x_i = \sum_{i=1}^{N} x'_i$. We conclude that $x_1 = x'_1$ and hence $x_i = x'_i$ for all $1 \leq i \leq N$. This proves that Φ is one to one. A slight modification of the proof of Lemma 3.8 shows that $\Phi : \mathbb{R}^{2N} \to \mathbb{R}^{2N}$ is a diffeomorphism. Finally the commutation relations (3.182) show that Φ is symplectic from $(\mathbb{R}^{2N}, \sum dx_i \wedge dy_i)$ onto $(\mathbb{R}^{2N}, \sum d\theta_i \wedge d\tilde{\lambda}_i)$. This completes the proof of Theorem 3.46. □

It remains to prove (3.167) for the θ_k.

Proof of (3.167) Recall from (3.76) that as $t \to \infty$, and $1 \leq k \leq N-1$,

$$x_k(t) - x_N(t) = 2t(\lambda_N - \lambda_k) + (N-k)\ln 4$$

$$+ \ln\left(\left(\frac{u_N(1)}{u_k(1)}\right)^2 \frac{\prod_{l=1}^{N-1}(\lambda_l - \lambda_N)^2}{\prod_{l=1}^{k-1}(\lambda_l - \lambda_k)^2}\right) + \text{e.s.e.} \quad (3.186)$$

Now as the Toda flow $(x(0), y(0)) \mapsto (x(t), y(t))$ is symplectic, it follows that

$$\{x_j(t), x_k(t)\} = \{x_j(0), x_k(0)\} = 0, \quad 1 \leq j, k \leq N,$$

for all t. Furthermore, it is easily seen from the proof of (3.186) that it can be differentiated term by term with respect to $x_j(0), y_k(0)$. It follows in particular that for $1 \leq j, k \leq N-1$,

$$0 = \{x_k(t) - x_N(t), x_j(t) - x_N(t)\} = \{2t(\lambda_N - \lambda_k), 2t(\lambda_N - \lambda_j)\}$$

$$+ \left\{2t(\lambda_N - \lambda_k), \ln\left(\left(\frac{u_N(1)}{u_j(1)}\right)^2 \frac{\prod_{l=1}^{N-1}(\lambda_l - \lambda_N)^2}{\prod_{l=1}^{j-1}(\lambda_l - \lambda_j)^2}\right)\right\}$$

$$+ \left\{\ln\left(\left(\frac{u_N(1)}{u_k(1)}\right)^2 \frac{\prod_{l=1}^{N-1}(\lambda_l - \lambda_N)^2}{\prod_{l=1}^{k-1}(\lambda_l - \lambda_k)^2}\right), 2t(\lambda_N - \lambda_j)\right\}$$

$$+ \left\{\ln\left(\left(\frac{u_N(1)}{u_k(1)}\right)^2 \frac{\prod_{l=1}^{N-1}(\lambda_l - \lambda_N)^2}{\prod_{l=1}^{k-1}(\lambda_l - \lambda_k)^2}\right), \ln\left(\left(\frac{u_N(1)}{u_j(1)}\right)^2 \frac{\prod_{l=1}^{N-1}(\lambda_l - \lambda_N)^2}{\prod_{l=1}^{j-1}(\lambda_l - \lambda_j)^2}\right)\right\}$$

$$+ \text{e.s.e.} \quad (3.187)$$

Matching terms of the same order in t, we conclude in particular that

$$\{\gamma_j, \gamma_k\} = 0, \quad 1 \leq j, k \leq N-1, \quad (3.188)$$

where

$$\gamma_k = \ln\left(\left(\frac{u_k(1)}{u_N(1)}\right)^2 \frac{\prod_{l=1}^{k-1}(\lambda_l - \lambda_k)^2}{\prod_{l=1}^{N-1}(\lambda_l - \lambda_N)^2}\right). \quad (3.189)$$

As in (3.157), $\nu_k = \ln \frac{u_k(1)}{u_N(1)}$. Let

$$\eta_k = 2\theta_k = \ln\left(\left(\frac{u_k(1)}{u_N(1)}\right)^2 \left|\frac{P'(\lambda_k)}{P'(\lambda_N)}\right|\right)$$

$$= 2\nu_k + \sum_{l \neq k} \ln|\lambda_k - \lambda_l| - \sum_{l \neq N} \ln|\lambda_N - \lambda_l| \qquad (3.190)$$

for $1 \leq k \leq N-1$. Now for $k < j$, using (3.160), we find

$$\{\gamma_k, \gamma_j\} = \{2\nu_k, 2\nu_j\} + \left\{2\nu_k, 2\sum_{l=1}^{j-1} \ln|\lambda_j - \lambda_l| - 2\sum_{l=1}^{N-1} \ln|\lambda_N - \lambda_l|\right\}$$

$$+ \left\{2\sum_{l=1}^{k-1} \ln|\lambda_k - \lambda_l| - 2\sum_{l=1}^{N-1} \ln|\lambda_N - \lambda_l|, 2\nu_j\right\}$$

$$= \{2\nu_k, 2\nu_j\} + \frac{4}{\lambda_k - \lambda_j} - \frac{4}{\lambda_k - \lambda_N}$$

$$- 4\sum_{l=1}^{N-1} \frac{1}{\lambda_l - \lambda_N} + \frac{4}{\lambda_j - \lambda_N} + 4\sum_{l=1}^{N-1} \frac{1}{\lambda_l - \lambda_N}$$

$$= \{2\nu_k, 2\nu_j\} + \frac{4}{\lambda_k - \lambda_j} - \frac{4}{\lambda_k - \lambda_N} + \frac{4}{\lambda_j - \lambda_N}. \qquad (3.191)$$

On the other hand, again for $k < j$,

$$\{\eta_k, \eta_j\} = \{2\nu_k, 2\nu_j\} + \left\{2\nu_k, \sum_{l \neq j} \ln|\lambda_j - \lambda_l| - \sum_{l \neq N} \ln|\lambda_N - \lambda_l|\right\}$$

$$+ \left\{\sum_{l \neq k} \ln|\lambda_k - \lambda_l| - \sum_{l \neq N} \ln|\lambda_N - \lambda_l|, 2\nu_j\right\}$$

$$= \{2\nu_k, 2\nu_j\} - \frac{2}{\lambda_j - \lambda_k} + \frac{2}{\lambda_j - \lambda_N} + \frac{2}{\lambda_N - \lambda_k} + 2\sum_{l \neq N} \frac{1}{\lambda_N - \lambda_l}$$

$$+ \frac{2}{\lambda_k - \lambda_j} - \frac{2}{\lambda_k - \lambda_N} - \frac{2}{\lambda_N - \lambda_j} - 2\sum_{l \neq N} \frac{1}{\lambda_N - \lambda_l}$$

$$= \{2\nu_k, 2\nu_j\} + \frac{4}{\lambda_k - \lambda_j} + \frac{4}{\lambda_N - \lambda_k} + \frac{4}{\lambda_j - \lambda_N}. \qquad (3.192)$$

From (3.191) and (3.192) we see that

$$\{\gamma_k, \gamma_j\} = 0 \Leftrightarrow \{\eta_k, \eta_j\} = 0,$$

and so

$$\{\theta_k, \theta_j\} = 0$$

by (3.188), as desired. □

Remark 3.47 Let

$$\hat{\theta}_k = \frac{1}{2}\gamma_k = \ln\left(\frac{u_k(1)}{u_N(1)}\frac{\prod_{l=1}^{k-1}|\lambda_l - \lambda_k|}{\prod_{l=1}^{N-1}|\lambda_l - \lambda_N|}\right), \qquad 1 \le k \le N-1, \quad (3.193)$$

$$\hat{\theta}_0 = \theta_0 = x_{cm}, \qquad (3.194)$$

with $\tilde{\lambda}_i = \lambda_i - \frac{1}{N}\sum_{j=1}^{N}\lambda_j, 1 \le j \le N-1$, and $\tilde{\lambda}_0 = y_T$ as in Theorem (3.46). It is straightforward to check that $\hat{\theta}_l, \tilde{\lambda}_p, 0 \le l, p \le N-1$ give a second set of canonical coordinates for $\left(\mathbb{R}^{2n}, \sum_{i=1}^{N} dx_i \wedge dy_i\right)$, i.e.,

$$\begin{aligned}\{\hat{\theta}_i, \hat{\theta}_j\} &= 0, & 0 \le i, j \le N-1, \\ \{\tilde{\lambda}_i, \tilde{\lambda}_j\} &= 0, & 0 \le i, j \le N-1, \\ \{\hat{\theta}_i, \tilde{\lambda}_j\} &= \delta_{ij}, & 0 \le i, j \le N-1.\end{aligned} \qquad (3.195)$$

Now it is an Exercise to show generally that if $x_i, y_j, 0 \le i, j \le N-1$ is a set of canonical variables, and $\hat{x}_i, y_j, 0 \le i, j \le N-1$ is a second set of canonical variables with the same actions y_j, then $\hat{x}_i - x_i$ depends only on y:

$$\hat{x}_i - x_i = f_i(y), \qquad 0 \le i \le N-1,$$

where

$$\frac{\partial f_i}{\partial y_j}(y) = \frac{\partial f_j}{\partial y_i}(y), \qquad 0 \le i, j \le N-1. \quad (3.196)$$

Conversely, if x_i, y_i are canonical variables and $\hat{x}_i = x_i + f_i(y)$ where f_i satisfies (3.196), then $\hat{x}_i, y_j$ is another set of canonical variables.

Exercise 3.48 Verify directly that $\hat{\theta}_i - \theta_i = f_i(\tilde{\lambda})$ for some functions f_i, and that

$$\frac{\partial f_i}{\partial \tilde{\lambda}_j}(\tilde{\lambda}) = \frac{\partial f_j}{\partial \tilde{\lambda}_i}(\tilde{\lambda}), \qquad 0 \le i, j \le N-1.$$

Exercise 3.49 Use (3.77) to give an independent proof of Theorem 3.17, namely, that

$$\{\lambda_k, \lambda_j\} = 0, \qquad 1 \le k, j \le N.$$

3.9 Action-Angle Variables for Toda Flow on Full Symmetric Matrices

Theorem 3.46 asserts, once we remove x_{cm} and y_T, that the Toda lattice is completely integrable on the lowest dimensional nontrivial coadjoint orbit $\mathcal{O}_c$, $c = $ const., of the lower triangular group Lo on its dual Lie algebra ℓ^*,

$$\mathcal{O}_c = \{Y \in \mathcal{J}_N : \operatorname{tr} Y = c\}, \quad \dim \mathcal{O}_c = 2N - 2.$$

It turns out that the Toda lattice is also completely integrable on generic orbits $\mathcal{O}$ of full real, symmetric matrices X. Such orbits have dimension $2\lfloor N^2/4 \rfloor$ and are constructed as follows (we follow Deift et al., 1986). For an $N \times N$ matrix $X = X_{ij}$, let

$$X_k = \{X_{ij} : k+1 \leq i \leq N, 1 \leq j \leq N-k\}, \quad 0 \leq k \leq \left\lfloor \frac{N}{2} \right\rfloor. \tag{3.197}$$

Thus X_k is the $(N-k) \times (N-k)$ matrix obtained from X by deleting the first k rows and last k columns. For $0 \leq k \leq \left\lfloor \frac{N}{2} \right\rfloor$, let

$$P_k(X, \lambda) = \det(X - \lambda)_k = \sum_{r=0}^{N-2k} E_{r,k}(X) \lambda^{N-2k-r}.$$

We say that a real, symmetric matrix X is *generic* if

$$E_{0,k}(X) \neq 0, \quad 0 \leq k \leq \left\lfloor \frac{N}{2} \right\rfloor. \tag{3.198}$$

If X is generic, then $P_k(X, \lambda)$ is exactly of degree $N - 2k$. It turns out that

$$\left\{ \frac{E_{rk}(X)}{E_{0,k}(X)} : 0 \leq k \leq \left\lfloor \frac{N-1}{2} \right\rfloor, \, 1 \leq r \leq N - 2k \right\},$$

gives rise to the commuting integrals for the Toda flow. Equivalently the generalized eigenvalues $\{\lambda_{jk}\}$ and generalized (right) eigenvectors $v_{r,k}$, for which

$$(X - \lambda_{r,k})_k v_{r,k} = 0, \quad v_{r,k} \neq 0, \quad 0 \leq k \leq \left\lfloor \frac{N-1}{2} \right\rfloor, 1 \leq r \leq N - 2k, \tag{3.199}$$

give rise to the commuting integrals, or actions, for the flow. Note that $\{\lambda_{r,0}\}_{r=1}^N$ are just the eigenvalues of X. Also, it turns out that the first components of the (suitably normalized) eigenvectors $v_{r,k}$ provide the angles for the flow.

For a generic X, it turns out that for $0 \leq k \leq \left\lfloor \frac{N}{2} \right\rfloor$ and for $0 \leq k' \leq \left\lfloor \frac{N-1}{2} \right\rfloor$, $E_{0,k}(X)$ does not change sign and $\dfrac{E_{1,k'}(X)}{E_{0,k'}(X)}$ is invariant under the coadjoint

3.9 Action-Angle Variables for Toda on Full Matrices

action of Lo on its dual Lie algebra l^*, and the coadjoint orbit $\mathcal{O}_X \subset l^*$ through X has the form

$$\mathcal{O}_X = \left\{ Y = Y^T \ : \ \operatorname{sgn} E_{0,k}(Y) = \operatorname{sgn} E_{0,k}(X), \quad 1 \leq k \leq \left\lfloor \frac{N}{2} \right\rfloor \right.$$

$$\left. \frac{E_{1k}(Y)}{E_{0,k}(Y)} = \frac{E_{1k}(X)}{E_{0,k}(X)}, \quad 0 \leq k \leq \left\lfloor \frac{N-1}{2} \right\rfloor \right\},$$

and

$$\dim \mathcal{O}_X = 2 \left\lfloor \frac{N^2}{4} \right\rfloor. \tag{3.200}$$

Note that it N is even, say $N = 2p$, then $\left\lfloor \frac{N-1}{2} \right\rfloor = p - 1$, and so

$$\dim \mathcal{O}_X = \frac{N(N+1)}{2} - p = p(2p+1) - p = 2p^2,$$

which is $2 \left\lfloor \frac{N^2}{4} \right\rfloor$. And if N is odd, say $N = 2p + 1$, then $\left\lfloor \frac{N-1}{2} \right\rfloor = p$ and

$$\dim \mathcal{O}_X = (2p+1)(p+1) - (p+1) = 2p(p+1),$$

which again is $2 \left\lfloor \frac{N^2}{4} \right\rfloor$. From (3.199) we see that there are

$$\sum_{k=0}^{\lfloor \frac{N-1}{2} \rfloor} (N - 2k),$$

generalized eigenvalues $\{\lambda_{r,k}\}$. For $N = 2p$ even, this sum is

$$\sum_{k=0}^{p-1}(2p - 2k) = 2p^2 - p(p-1) = p(p+1).$$

Also, for each $0 \leq k \leq \left\lfloor \frac{N-1}{2} \right\rfloor$, specifying the traces

$$\sum_{r=1}^{N-2k} \lambda_{r,k} = -\frac{E_{1k}}{E_{0,k}}$$

is clearly equivalent to specifying the co-adjoint invariants $E_{1k}/E_{0,k}$. Hence the number of generalized eigenvalues, modulo the traces $\sum_{r=1}^{N-2k} \lambda_{r,k}$, is

$$p(p+1) - \left(\left\lfloor \frac{N-1}{2} \right\rfloor + 1 \right) = p(p+1) - p = p^2 = \left\lfloor \frac{N^2}{4} \right\rfloor,$$

so the count is correct. The case when N is odd is similar.

To illustrate the situation, for $N = 4$, $X = (X_{ij})_{1 \leq i,j \leq 4}$ and $\lfloor \frac{(N-1)}{2} \rfloor = 1$

$$(X - \lambda)_1 = \begin{pmatrix} X_{21} & X_{22} - \lambda & X_{23} \\ X_{31} & X_{32} & X_{33} - \lambda \\ X_{41} & X_{42} & X_{43} \end{pmatrix},$$

and so for $E_{01} = X_{41} \neq 0$, $P_1(X, \lambda)$ has two roots, $\lambda_{01}, \lambda_{02}$, with trace $\lambda_{01} + \lambda_{02}$. For $k \neq 0$, $P_0(X, \lambda) = \det(X - \lambda)$ has four roots $\lambda_{01}, \lambda_{02}, \lambda_{03}, \lambda_{04}$, the standard eigenvalues of X with standard trace $\lambda_{01} + \lambda_{02} + \lambda_{03} + \lambda_{04}$. Thus, modulo the traces, we see that there are $1 + 3 = 4$ generalized eigenvalues, which matches $\lfloor N^2/4 \rfloor = 4$. For $N = 5$, with $\lfloor \frac{(N-1)}{2} \rfloor = 2$,

$$(X - \lambda)_1 = \begin{pmatrix} X_{21} & X_{22} - \lambda & X_{23} & X_{24} \\ X_{31} & X_{32} & X_{33} - \lambda & X_{34} \\ X_{41} & X_{42} & X_{43} & X_{44} - \lambda \\ X_{51} & X_{52} & X_{53} & X_{54} \end{pmatrix},$$

and so for $E_{01} = -X_{51} \neq 0$, $P_1(X, \lambda)$ has three roots $\lambda_{11}, \lambda_{12}, \lambda_{13}$, with trace $\lambda_{11} + \lambda_{12} + \lambda_{13}$. For $k = 2$,

$$(X - \lambda)_2 = \begin{pmatrix} X_{31} & X_{32} & X_{33} - \lambda \\ X_{41} & X_{42} & X_{43} \\ X_{51} & X_{52} & X_{53} \end{pmatrix},$$

and so for

$$E_{02} = -\det \begin{pmatrix} X_{41} & X_{42} \\ X_{51} & X_{52} \end{pmatrix},$$

$P_2(X, \lambda)$ has one root λ_{23}, which is also the trace. Together with the eigenvalues, $\lambda_{01}, \lambda_{02}, \lambda_{03}, \lambda_{04}, \lambda_{05}$, we see that modulo the traces we have $2 + 0 + 4 = 6$, which matches $\lfloor 5^2/4 \rfloor$.

What are the angles corresponding to the $\lambda_{r,k}$?

For $k = 0$, we anticipate that the ratios

$$\mu_{r,0} = \frac{1}{2} \ln \left| \frac{D(M, \lambda_{r,0})}{D(M, \lambda_{N,0})} \right|, \quad 1 \leq r \leq N - 1,$$

(see (3.171)) will provide the angles conjugate to the $\lambda_{r,0}$. The point to note is that the $D(M, \lambda_{05})$ reflect the basic analytical fact that the eigenvectors $\{u_j\}$ of a real symmetric matrix $X = X^T$ have a natural normalization, viz. $\sum_{i=1}^N (u_{ij})^2 = 1$, so that u_{1j}^2 is well-defined, $1 \leq j \leq N$. But for $k > 0$ we are dealing with a generalized eigenvalue problem, in fact two generalized eigenvalue problems, a right generalized eigenvalue problem

$$(X - \lambda_{r,k})v_{r,k} = 0,$$

3.9 Action-Angle Variables for Toda on Full Matrices

and a left generalized eigenvalue problem

$$\hat{v}_{r,k}(X - \lambda_{r,k}) = 0,$$

and there is no a priori general way to normalize $v_{r,k}$ or $\hat{v}_{r,k}$. In particular, the components $v_{r,k}(i)$ or $\hat{v}_{r,k}(i)$, that we may anticipate to give rise to angles conjugate to $\lambda_{r,k}$, are not a priori well-defined. However, quite remarkably, there is a natural way to specify $v_{r,k}$ and $\hat{v}_{r,k}$.

Consider the case $N = 5$ and set

$$Q_{51}(\lambda) = \langle e_5, \frac{1}{X - \lambda} e_1 \rangle = \langle \frac{1}{X - \lambda} e_5, e_1 \rangle.$$

By Cramer's rule,

$$Q_{51}(\lambda) = \frac{\det(X - \lambda)_1}{\det(X - \lambda)},$$

and so $Q_{51}(\lambda_{r1}) = 0$ for any generalized eigenvalue λ_{r1} of $(X)_1$. It follows that $v_{r1} = (X - \lambda_{r1})^{-1} e_1$ has the form $(w_{r1}, 0)^T$ for some $w_{r1} \in \mathbb{C}^4$. The relation $(X - \lambda_{r1}) v_{r1} = e_1$ then implies that $(X - \lambda_{r1})_1 w_{r1} = 0$, so that $w_{r1} \neq 0$ is a right generalized eigenvector for X_1. A similar calculation shows that $\hat{v}_{r1} = (X - \lambda_{r1})^{-1} e_5$ has the form $(0, \hat{w}_{r1})^T$ for some $\hat{w}_{r1} \in \mathbb{C}^4$ and $\hat{w}_{r1}^T (X - \lambda_{r1}) = 0$, so that $\hat{w}_{r1}^T$ is a left generalized eigenvector for X_1. Let

$$f_{r1}(X) = \hat{w}_{r1}(1) = \langle e_2, (X - \lambda_{r1})^{-1} e_5 \rangle,$$

and set

$$\mu_{r1} = \frac{f_{r1}(X)}{f_{31}(X)}, 1 \leq r \leq 2.$$

It turns out that the $\ln \mu_{r,1}$ are conjugate to the λ_{r1}, $1 \leq r \leq 2$. Note that μ_{r1} is the first component of the suitably normalized generalized left eigenvector $\hat{w}_{r1}$. On the other hand, the quantities

$$w_{r1}(4) = \langle e_4, (X - \lambda_{r1})^{-1} e_1 \rangle$$

give rise to angles conjugate to a different set of actions λ_{s1}, $2 \leq s \leq 3$. (Recall that $\lambda_{11} + \lambda_{21} + \lambda_{31}$ is a co-adjoint invariant for the Toda system, and so we can choose, for example $\lambda_{11}, \lambda_{21}$ or $\lambda_{21}, \lambda_{31}$ for the actions). The above construction of v_{rk} and $\hat{v}_{rk}$ generalizes for $k > 0$ leading, in particular, to the analogs $f_{rk}(X)$ (resp. $\mu_{rk}(X)$) of $f_{r1}(X)$ (resp. $\mu_{r1}(X)$). In this way, one constructs actions $\{\lambda_{r,k}\}$ and angles $\{\ln \mu_{r,k}\}$ for the Toda flow, with

$$0 \leq k \leq \left\lfloor \frac{N-1}{2} \right\rfloor, \quad 1 \leq r \leq N - 2k - 1, \quad \sum_{k=0}^{\lfloor (N-1)/2 \rfloor} (N - 2k - 1) = \left\lfloor \frac{N^2}{4} \right\rfloor,$$

so proving that the Toda flow on generic orbits of dimension $2\lfloor N^2/4 \rfloor$ is completely integrable (see Deift et al., 1986). Moreover, under the Toda flow all the $\{\lambda_{r,k}\}$ and $\{\ln \mu_{r,k}\}$ are constant, except for $\ln \mu_{r,0}$, which evolves linearly:

$$\ln \mu_{r,0}(t) = \frac{1}{2} \ln \left| \frac{D(X(t), \lambda_{r,0})}{D(X(t), \lambda_{N,0})} \right| = \ln |\mu_{r,0}(0)| + (\lambda_{r,0} - \lambda_{rN})t, \quad 1 \le r \le N-1,$$

(cf. (3.184)), noting that $\lambda_i - \lambda_N = \tilde{\lambda}_i + \frac{1}{N} \sum_{j=1}^{N} \tilde{\lambda}_j$, where $\lambda_i = \lambda_{i0}$. In the former case, we use the Hamiltonian $\frac{1}{2} \operatorname{tr} X^2$ rather than $H_{\text{Toda}} = 2 \operatorname{tr} X^2$.

3.10 The Stroboscope Theorem

The stroboscope property in Theorem 1.2 for real positive matrices follows from Theorem 3.20 applied to the function $f(s) = -(s \ln s - s)$. Here $\ln s$ is specified to be analytic in the complement of $\mathbb{R}_-$ in $\mathbb{C}$, and real on $\mathbb{R}_+$. The associated Hamiltonian $H_{QR}(X) = \operatorname{tr} f(X)$ leads to the flow (1.23) in Theorem 1.2 (recall that $\pi_k Y = -B(Y)$ for real symmetric Y),

$$\partial_t X = [X, \pi_k f'(X)] = [X, B(\ln(X))], \quad X(0) = X_0 = X_0^T. \quad (3.201)$$

It then follows from Corollary 3.34 that

$$X(n) = e^{\ln X(n)}$$

is obtained by performing n QR steps on $e^{\ln X_0} = X_0$, i.e.,

$$X(n) = \phi_{QR}^n(X_0)$$

as in (1.24). Finally, as

$$H_{QR}(X) = -\sum_{k=1}^{N} (\lambda_k \ln \lambda_k - \lambda_k),$$

where $\{\lambda_k\}_{k=1}^{N}$ are the eigenvalues of X, it is clear that H_{QR} generates a completely integrable flow on an invariant subset of the Poisson manifold of real symmetric matrices. Moreover, the flow $X(t)$ commutes with the Toda flow generated by (1.1). This completes the proof of Theorem 1.2.

In order to accommodate matrices which are real symmetric but not necessarily positive definite, we combine equations (1.22) and (3.121) and consider the following generalized QR flow $t \mapsto X(t)$ on Hermitian matrices,

$$\partial_t X = [X, \tilde{B}(\ln(X))], \quad X(0) = X_0 = X_0^* \in \Gamma_N. \quad (3.202)$$

3.10 The Stroboscope Theorem

Here the matrix $\ln(X(t))$ is defined as in (3.113), where $f(s) = \ln s$ is real on $\mathbb{R}_+$ and the contour $\mathcal{C}$ lies, for example, in the complement of the negative imaginary numbers in $\mathbb{C}$, and the eigenvalues λ_k of X lie in the interior of $\mathcal{C}$. Again Corollary 3.34 applies and again we have the stroboscope property

$$X(n) = \phi_{QR}^n(X_0),$$

$$f(X) := \frac{1}{2i\pi} \int_{\mathcal{C}} \frac{f(s)}{s - X} ds. \tag{3.203}$$

Remark 3.50 If X_0 is not positive definite, (3.202) must be replaced by

$$\partial_t X = [X, \widehat{B}(\ln(X))], \quad X(0) = X_0 = X_0^* \in \Gamma_N, \tag{3.204}$$

where

$$\widehat{B}(Y) = Y_- - Y_-^* + i\mathfrak{J} \operatorname{diag}(Y) \tag{3.205}$$

for any matrix Y. Here $(i(\mathfrak{J} \operatorname{diag}(Y)))_{kk} = iy_k$, where $Y_{kk} = x_k + iy_k$. With this redefinition, $\widetilde{B} \mapsto \widehat{B}$, solutions $X(t)$ of (3.204) can be computed via Symes factorization as before, $X(t) = Q(t)^* X_0 Q(t)$, where

$$e^{t \ln X_0} = Q(t) R(t)$$

is the standard QR factorization of $e^{t \ln X_0}$. For general $t > 0$, $X(t)$ may not be real, but for $t = k$ for any positive integer k, $e^{k \ln X_0} = X_0^k = Q(k)R(k)$ and so $Q(k)$ as well as $X(k) = Q(k)^T X_0 Q(k)$ are real. Again we have the stroboscope property, $X(k) = \phi_{QR}^k(X_0)$. Also, the vector field (3.204) is Hamiltonian with respect to the Lie-Poisson structure associated with the Lie group of lower triangular matrices with complex entries and positive diagonal elements. We leave the details to the interested reader.

Suppose $X(t)$ is the solution of the QR flow (1.22) with $X(0) = X_0$, where X_0 is real and positive definite. Then we apply the method in the proof of Theorem 3.9 to conclude that

$$u_{1j}(t) = \frac{u_{1j}(0)\lambda_j^t}{\left(\sum_{i=1}^N u_{ij}^2(0)\lambda_i^{2t}\right)^{1/2}}, \quad 1 \leq j \leq N, \tag{3.206}$$

where $\{u_{1j}(t)\}$ are the first components of the normalized eigenvectors of $X(t)$. As noted above, if X_0 has negative eigenvalues, $X(t) = X(t)^T$ may not be real valued for all $t > 0$: However, the iterates $X_0, X(1) = X_1, \ldots, X(k) = X_k$ starting from the (invertible) matrix $X_0 \in \mathfrak{lo}^*$ are real symmetric. Let

u_k be a normalized eigenvector for X_0, $X_0 u_k = \lambda_k u_k$. By the QR iteration procedure,

$$X_1 = R_0 Q_0 = Q_0^T X_0 Q_0,$$

where $X_0 = Q_0 R_0$, and so

$$u_k(1) = Q_0^T u_k$$

is a normalized eigenvector for X_1 corresponding to the eigenvalue λ_k. Now, using the fact that R_0 is triangular,

$$\begin{aligned} u_{1k}(1) &= \langle e_1, u_k(1)\rangle = \langle e_1, Q_0^T u_k\rangle \\ &= \langle e_1, R_0^{-T} X_0 u_k\rangle = \lambda_k \langle R_0^{-1} e_1, u_k\rangle \\ &= \frac{\lambda_k}{(R_0)_{11}} u_{1k}. \end{aligned}$$

Hence,

$$\frac{u_{1k}(1)}{u_{1l}(1)} = \frac{\lambda_k}{\lambda_l} \frac{u_{1k}}{u_{1l}}, \qquad 1 \le k, l \le N, \tag{3.207}$$

which implies

$$u_{1j}(1) = \frac{u_{1j}\lambda_j}{\left(\sum_{i=1}^{N} u_{1,i}^2 \lambda_i^2\right)^{1/2}}, \qquad 1 \le j \le N, \tag{3.208}$$

(cf. (3.206)): Note that (3.206) is true for all $t > 0$ in the case where X_0 is real and positive definite, whereas (3.208) is true only for integer times $t = 1, 2, \ldots$, but now for any invertible real matrix. Iterating (3.208) we see that

$$u_{1j}(k) = \frac{u_{1j}\lambda_j^k}{\left(\sum_{i=1}^{N} u_{1,i}^2 \lambda_i^{2k}\right)^{1/2}}, \qquad 1 \le k, j \le N, \tag{3.209}$$

where $u_{1j}(k)$ is the first component of the eigenvector $u_j(k)$ of X_k.

Finally we note that the Toda flow

$$\partial_t X = [X, B(X)], \qquad X(0) = X_0,$$

exists globally for all real (*not necessarily symmetric*) $N \times N$ matrices X_0, and again gives rise to a completely integrable system, together with a stroboscope theorem on generic orbits. Again, the underlying symplectic structure arises from the action of a Lie group on its dual Lie algebra. The reader is referred to Deift et al. (1980) for more details.

Remark 3.51 As we see, the Toda flow $t \mapsto X(t)$ has many conserved quantities which are non-transcendental and explicitly computable. It is a tantalizing and open question how to use these conserved quantities to enhance eigenvalue computation.

4
Toda without Hamiltonian Structure

There are still algebraic surprises lying concealed within the Toda flow that we have not yet described. A simple argument with matrix factorizations gives rise to new inverse variables, the Z-coordinates, together with new commuting vector fields.

In these new variables, the Toda flows become explicit, straight line motions in $\mathbb{R}^N$ for an appropriate dimension N (Toda flows on full matrices are also considered). Moreover, orbit limits, such as diagonal matrices, lie beyond the scope of standard variables (see, e.g., M below for Jacobi matrices), and the asymptotic analysis must proceed using ad hoc methods. Such orbit limits, however, belong to the domain of the new variables, and asymptotic computations are easily performed through local theory. As described in Remark 4.14, the methods in this section extend the purview of the Toda system substantially.

The results in this chapter go back to Leite et al. (2008). Recent developments can be found in Torres and Tomei (2013) and Leite et al. (2023).

4.1 Another Set of Linearizing Variables

For a vector $v \in \mathbb{R}^k$, denote its entries by v_i. Proposition 3.8 provides coordinates on the set $\mathcal{J}$ of $N \times N$ Jacobi matrices. We repeat its statement with small notational changes:

There is a diffeomorphism between the set of $N \times N$ Jacobi matrices $\mathcal{J}$ and

$$M = \left\{ (\lambda, q) \in \mathbb{R}^N \times \mathbb{R}^N \mid \lambda_1 > \cdots > \lambda_N, \sum_1^N q_i^2 = 1, q_i > 0 \right\}.$$

4.1 Another Set of Linearizing Variables

For $X \in \mathcal{J}$, λ is the vector whose entries are the eigenvalues of X in strictly decreasing order, and q are the first coordinates of the associated appropriately normalized eigenvectors.

We introduce some notation to describe an alternative coordinatization of $\mathcal{J}$ and of many other sets. All matrices in this section are real, of dimension n.

The lower triangular group Lo and its Lie algebra $\mathfrak{lo}$ were defined in Section 3.4. Let $Lo^1 \subset Lo$ be the subgroup of matrices with diagonal entries equal to 1, with Lie algebra $\mathfrak{lo}^0$, the vector space of lower triangular matrices with null diagonal. We denote the group of (invertible) upper triangular matrices by Up, and by $Up^+ \subset Up$ the subgroup of matrices with strictly positive diagonal entries.

An *LU factorization* of an invertible matrix M is a factorization

$$M = LU, \quad L \in Lo^1, \quad U \in Up.$$

If this decomposition exists, it is unique. The *principal minors* of M are the determinants of the matrices with entries in the first k rows and columns of M. We say an invertible matrix is LU-positive if it admits an LU factorization.

Exercise 4.1 Show that a matrix M is LU-positive if and only if its principal minors are nonzero and then this decomposition is unique. Moreover, given an invertible matrix M, there is a permutation matrix P for which PM is LU-positive.

An invertible matrix M admits an *LU-positive decomposition* if

$$M = LU, \quad L \in Lo^1, \quad U \in Up^+.$$

Let S_N be the permutation group in n symbols, or, more concretely, the set of bijections of the set $\{1, \ldots, n\}$ to itself. For $\lambda_1 > \cdots > \lambda_N$ and $\pi \in S_N$, define

$$\Lambda^\pi = \mathrm{diag}(\lambda_{\pi(1)}, \ldots, \lambda_{\pi(N)}) = \mathrm{diag}(\lambda_1^\pi, \ldots, \lambda_N^\pi).$$

The permutation π induces another ordering of the originally strictly decreasing numbers λ_i. The second notation will allow us to drop the permutation in some arguments. Denote by D^π the cone of diagonal matrices with distinct diagonal entries ordered according to π,

$$D^\pi = \{\mathrm{diag}(d_1, \ldots, d_N) \mid d_i \in \mathbb{R}, d_{\pi(1)} > \cdots > d_{\pi(N)}\}.$$

Recall that $\Sigma = \Sigma_N$ denotes the vector space of real, $N \times N$ symmetric matrices. Let $\Sigma^s \subset \Sigma$ be the open set of real, symmetric matrices with distinct eigenvalues.

Exercise 4.2 Use results from Theorem 3.6 to show the following.

1. Eigenvalues of matrices in Σ^s may be ordered so that $\lambda_1 > \cdots > \lambda_N$. Use perturbation theory to show that there is a smooth, in fact real analytic, spectral decomposition $S = Q^T \Lambda Q$, with $Q = Q(S) \in SO$, the group of orthogonal matrices of determinant equal to 1.
2. Use the connectivity of $SO(N)$ to show that Σ^s is connected.

For $\pi \in S_N$, a matrix $S \in \Sigma^s$ is π-admissible if $S = Q^T \Lambda^\pi Q$, for some diagonal matrix $\Lambda^\pi \in D^\pi$ and an orthogonal matrix Q admitting an LU-positive decomposition. Finally set

$$\Sigma^\pi = \{S \in \Sigma^s \mid S \text{ is } \pi\text{-admissible}\}.$$

A matrix E is *sign diagonal* if it is real, diagonal, and $E^2 = I$.

How abundant are π-admissible matrices?

Lemma 4.3 *The following holds.*

(1) *A diagonal matrix D with simple spectrum is π-admissible only for $\pi = e$, the trivial permutation.*
(2) *Let $S \in \Sigma^s$, $S = \tilde{Q}^T \Lambda^\pi \tilde{Q}$, for $\Lambda^\pi \in D^\pi$ and an orthogonal matrix $\tilde{Q}$ admitting an LU decomposition. Then $S \in \Sigma^\pi$.*
(3) *For each $\pi \in S_N$, Σ^π is an open, dense subset of Σ^s.*
(4) *Every $S \in \Sigma^s$ is π-admissible for some $\pi \in S_N$. Thus $\Sigma^s = \cup_{\pi \in S_N} \Sigma^\pi$.*
(5) *Jacobi matrices are π-admissible for every $\pi \in S_N$.*

Proof To prove (1), write $D = Q^T D^\pi Q$ for Q LU-positive. Then $D = \hat{Q}^T D \hat{Q}$, where $\hat{Q} = PQ$ for some permutation matrix P. As D has simple spectrum, it follows that $\hat{Q} = \Delta$ is a diagonal matrix with $\Delta^2 = I$, so that $\Delta_{ii} = \pm 1$. As $PQ\Delta = I$, we conclude that $Q\Delta$ is the permutation matrix P^{-1}. But the minors of $Q\Delta$ are nonzero, which implies that $P^{-1} = I$. It then follows that $P = I$, $\pi = e$, the identity permutation, and $D^\pi = D$.

For (2), given $S = \tilde{Q}^T \Lambda^\pi \tilde{Q}$ and a sign diagonal matrix E,

$$S = \tilde{Q}^T \Lambda^\pi \tilde{Q} = (E\tilde{Q})^* \Lambda^\pi (E\tilde{Q}).$$

Clearly, there is some E for which $Q = E\tilde{Q}$ admits an LU-positive decomposition. Thus S is indeed π-admissible.

4.1 Another Set of Linearizing Variables

We prove (3). For $\pi \in S_N$, Σ^π is an open set of Σ by continuity. For $S \in \Sigma^s$ and $S = Q^T \Lambda^\pi Q$, for $\Lambda^\pi \in D^\pi$, we have $S \in \Sigma^\pi$ if and only if each principal minor of Q is nonzero. The square of a minor is a real analytic map in the open, connected set Σ^s, and thus cannot be zero in a ball of Σ^s, unless it is zero for all matrices in Σ^s, which is not the case, as $S = \mathrm{diag}(1, 2, \ldots, n)$ lies in Σ^s. Using (2), we conclude that Σ^π is also dense in Σ.

For (4), diagonalize, $S = Q^T \Lambda Q$, with $\Lambda = \mathrm{diag}(\lambda_1 > \cdots > \lambda_N)$ and Q orthogonal. The rows or Q are (normalized) eigenvectors. As Q is invertible, there is a reordering of its rows, $\widetilde{Q} = PQ$ for some permutation matrix P such that the principal minors of $\widetilde{Q}$ are not zero. As before, there is a sign diagonal matrix E, for which $Q = E\widetilde{Q}$ admits an LU-positive decomposition.

For the last item, from the proof of Proposition 3.8,

$$V^\pi = [u \ \Lambda^\pi u \cdots (\Lambda^\pi)^{n-1} u] = Q^T R,$$

and as R is upper triangular and the principal minors of V^π are nonzero, it follows that the same is true for the principal minors of Q^T, and hence of Q. Thus the Jacobi matrices are π-admissible for every $\pi \in S_N$. □

Remark 4.4 Items (1) and (5) do not contradict each other: a real diagonal matrix is not Jacobi.

Fix $\pi \in S_N$. We define a diffeomorphism between Σ^π and $\mathcal{Z}^\pi$, where

$$\mathcal{Z}^\pi = \{\Lambda^\pi + T \mid \Lambda^\pi \in D^\pi, \ T \in \mathfrak{l}\mathfrak{o}^0\} \simeq D^\pi \times \mathfrak{l}\mathfrak{o}^0.$$

Take $Z \in \mathcal{Z}^\pi$. Its (simple) eigenvalues form a diagonal matrix $\Lambda^\pi = \mathrm{diag}\, Z \in D^\pi$. Diagonalize $Z = L^{-1} \Lambda^\pi L$, where, as above, the columns of $L^{-1} \in Lo^1$ consist of uniquely normalized eigenvectors of Z. The (unique) QR decomposition $L = QR$ obtains Q. As the minors of L^{-1} are positive, the same is true for the minors of L and $Q = LR^{-1}$ as $R \in Up^+$. Set $X = Q^T \Lambda^\pi Q$ and define $\phi^\pi(Z) = X \in \Sigma^\pi$.

Remark 4.5 If $S = Q^T \Lambda^\pi Q$ for some LU-positive Q, then Q is unique. Indeed, if $S = Q_0^T \Lambda^\pi Q_0$ for some other LU-positive Q_0, then clearly $Q_0^T = Q^T E$ for some sign diagonal matrix E. But as both Q and Q_0 are LU-positive, we must have $E = I$, and so $Q = Q_0$.

Theorem 4.6 *(inverse variables)* $\phi^\pi : \mathcal{Z}^\pi \to \Sigma^\pi$ *is a diffeomorphism.*

Proof The map is smooth: we construct its inverse. For $S \in \Sigma^\pi$, write $S = Q^T \Lambda^\pi Q$, for $\Lambda^\pi \in D^\pi$ and $Q = LU$ its LU-positive decomposition. (Note that Q is unique by Remark 4.5.) Set $Z = L^{-1}\Lambda^\pi L \in \mathcal{Z}^\pi$. The map $\psi : \Sigma^\pi \to \mathcal{Z}^\pi$, $S \mapsto Z$ is the inverse of ϕ. □

Let $\psi^\pi = (\phi^\pi)^{-1} : \Sigma^\pi \to \mathcal{Z}^\pi$: the matrices $Z = \psi^\pi(S)$ are the *Z-variables*. To write the Toda flow in Z-variables, we follow the notation in Theorem 3.15: for a continuous real-valued function f,

$$\partial_t X = [X, B(f(X))], \quad X(0) = X_0 \in \Sigma^s, \tag{4.1}$$

where, for $X = Q^T \Lambda Q$, $f(X) = Q^T \mathrm{diag}(f(\lambda_1), \ldots, f(\lambda_N))Q$. From Lemma 4.3 (2), $X_0 \in \Sigma^\pi$ for some $\pi \in S_N$.

Remark 4.7 For a real matrix $A = V\Lambda V^{-1}$ with real simple spectrum $\Lambda = \mathrm{diag}(\lambda_1, \ldots, \lambda_N)$, and eigenvector matrix V,

$$f(A) = V \, \mathrm{diag}(f(\lambda_1), \ldots, f(\lambda_N)) \, V^{-1}$$

defines $f(A)$ unambiguously for all continuous functions f defined on an interval $I \supset \sigma(A)$. For such A and any invertible real matrix W, we find

$$f(WAW^{-1}) = f(WV\Lambda(WV)^{-1})$$
$$= WV\mathrm{diag}(f(\lambda_1), \ldots, f(\lambda_N))(WV)^{-1} = Wf(A)W^{-1}.$$

Theorem 4.8 (Linearization) Let $X(t) \in \Sigma^s$ be the solution of equation (4.1) where $X_0 \in \Sigma^\pi$ for some $\pi \in S_N$. Then, for all $t \in \mathbb{R}$, $X(t) \in \Sigma^\pi$ and the curve $Z(t) = (\psi^\pi)(X(t)) \in \mathcal{Z}^\pi$ satisfies

$$\partial_t Z = [f(\Lambda^\pi), Z], \quad Z(0) = Z_0 = \psi^\pi(X_0). \tag{4.2}$$

In particular, $Z(t) = e^{tf(\Lambda^\pi)} Z_0 \, e^{-tf(\Lambda^\pi)}$: entries evolve as

$$Z_{ii}(t) = Z_{ii}(0) = \lambda_{\pi(i)}, \quad Z_{ij}(t) = e^{t(f(\lambda_{\pi(i)}) - f(\lambda_{\pi(j)}))} Z_{ij}(0). \tag{4.3}$$

Thus $\partial_t Z$ is a commutator of Z with a *constant* matrix.

Proof The proof will proceed as follows. From Theorem 4.6, $\phi^\pi : \mathcal{Z}^\pi \subset \mathfrak{L}_0 \to \Sigma^\pi \subset \Sigma^s$ is a diffeomorphism between open sets. The differential equation is associated with the vector field $\Psi(X) = [X, \pi_k f(X)] \in \Sigma$ and we have to show that its pullback under ϕ^π (i.e., changing to Z-variables) is

$$(\phi^\pi)^*(\Psi)(Z) = [Z, -f(\Lambda^\pi)] \in \mathfrak{L}_0. \tag{4.4}$$

The formula of the evolution of the entries of $Z(t)$ then implies that, for $Z(0) = Z_0 \in \mathcal{Z}^\pi$, we have $Z(t) \in \mathcal{Z}^\pi$ for all $t \in \mathbb{R}$. Said differently, $\mathcal{Z}^\pi$ is invariant under the flow induced by $\phi^*(\Psi)$, and thus Σ^π is invariant under the Toda vector field $\Psi: X(t) = \phi^\pi(Z(t))$ then is also defined for all $t \in \mathbb{R}$.

4.1 Another Set of Linearizing Variables

We now prove equation (4.4). From Theorem 1.1, for any $X_* \in \Sigma^\pi$, the solution $X(t)$ of the Toda flow with $X(0) = X_*$ is given by

$$X(t) = Q(t)^T \Lambda^\pi Q(t), \quad X(0) = X_* \in \Sigma^\pi.$$

As Σ^π is an open set, for t close to zero, the orthogonal matrix $Q(t)$ admits an LU-positive decomposition $Q(t) = L(t)U(t)$, $L(t) \in Lo^1$, $U(t) \in Up^+$. We drop the dependence on $t \in \mathbb{R}$ and $\pi \in S_N$. As in the beginning of Section 3.6,

$$\partial_t Q = Q \pi_k f(X), \quad Q(0) = Q_0,$$

where Q_0 is obtained from $X(0) = Q_0^T \Lambda Q_0$.

Consider two pairs of complementary projections: $\pi_{s\ell}$ and $\tilde{\pi}_u$, taking values on strictly lower and upper triangular matrices, respectively, and π_k and π_u, taking values on skew-symmetric matrices and upper triangular matrices. Notice that π_u and $\tilde{\pi}_u$ are **not** equal. More precisely, for any matrix $Y = Y_- + Y_0 + Y_+$, $\pi_{s\ell}(Y) = Y_-$, $\tilde{\pi}_u(Y) = Y_0 + Y_+$. and $\pi_k(Y) = Y_- - Y_-^T$, $\pi_u(Y) = Y_-^T + Y_0 + Y_+$.

Differentiating $Q(t) = L(t)U(t)$, we obtain $\partial_t Q = \partial_t L U + L \partial_t U$ so that

$$\begin{aligned} L^{-1}\partial_t L &= \pi_{s\ell}\left(L^{-1}(Q\pi_k f(X)) U^{-1}\right) = \pi_{s\ell}\left(U (\pi_k f(X)) U^{-1}\right) \\ &= \pi_{s\ell}\left(U f(X) U^{-1}\right) - \pi_{s\ell}\left(U (\pi_u f(X)) U^{-1}\right) = \pi_{s\ell}\left(f(UXU^{-1})\right), \end{aligned}$$

as the last second term is zero, since $U (\pi_u f(X)) U^{-1}$ is upper triangular. As $UXU^{-1} = UQ^{-1}\Lambda Q U^{-1} = L^{-1}\Lambda L = Z$, we obtain

$$L^{-1}\partial_t L = \pi_{s\ell}\left(f(Z)\right).$$

As the matrix $f(Z) = L^{-1}f(\Lambda)L$ is lower triangular with diagonal $f(\Lambda)$, we have

$$L^{-1}\partial_t L = f(Z) - f(\Lambda).$$

We now compute the evolution of $Z = L^{-1}\Lambda L$. Differentiating,

$$\partial_t Z = [Z, L^{-1}\partial_t L] = [Z, f(Z) - f(\Lambda)] = [Z, -f(\Lambda)] = [f(\Lambda), Z].$$

The upshot is that, locally, the map $Z = (\psi^\pi)(X) \in \mathcal{Z}^\pi$ converts solutions of the Toda flow for X into solutions of the evolution given by equation (4.2, and conversely. Finally, as $f(\Lambda)$ is a constant, equations (4.3) are immediate, and imply global existence for $Z(t) \in \mathcal{Z}^\pi$, hence for $X(t) \in \Sigma^\pi$. □

4.2 Profiles and Isospectral Manifolds

So far, we introduced Z-variables by the map $\psi^\pi = (\phi^\pi)^{-1} : \Sigma^\pi \to \mathcal{Z}^\pi$ on open, dense, sets Σ^π which cover Σ^s (the real, symmetric matrices with simple spectrum). The sets Σ^π are invariant under the Toda flow, and on $\mathcal{Z}^\pi$ the flow is linear. We now show that Z-variables restrict well on special vector spaces of matrices, by specifying a *profile*, to be defined below. In particular, this is how Z-variables are induced on tridiagonal symmetric matrices, so that they become convenient tools for the study of the original Toda lattice.

From Theorem 1.1, the solution to the Toda differential equation admits representations as conjugations of the initial condition by orthogonal or upper triangular matrices. This leads us to consider some vector spaces of matrices which are invariant under upper triangular conjugations.

We consider entries (i, j) of $N \times N$ matrices. For entry (i, j) in the lower triangular part of a matrix, so that $i \geq j$, we define the (i, j)-*quadrant* to be the set of entries (i', j') with $i' \leq i$ and $j' \geq j$. Given a set of entries P, the profile associated with P is the union of entries in the (i, j)-quadrants, for $(i, j) \in P$.

Exercise 4.9 Show that different entry sets may induce the same profile. We say that an entry set p is *minimal* if $\#p \leq \#P$ for all entry sets P that generate the same profile. Show that p is minimal if there are no two entries in p in the same row or in the same column. Show also that every set of entries P admits a unique minimal subset p with the same profile.

Consider the following example for 4×4 matrices: $C' = \{(3, 2), (2, 1)\}$ and $C = \{(3, 2), (2, 1), (2, 2)\}$. The set C' is minimal.

Let $E_{ij} = e_i e_j^T$ be the matrix with a single nonzero entry equal to 1 at (ij). Given a set of entries P with minimal set p, we denote by V_p the vector space of matrices generated by the matrices E_{ij} for which entry (ij) belongs to the profile associated with P (equivalently, by p).

A simple example is the set of *upper Hessenberg* matrices, all of whose entries $(i, j), i > j+1$, are zero. In this case, $p = \{(2, 1), (3, 2), \ldots, (n, n-1)\}$. The space $M(n, \mathbb{R})$ is V_p for $p = \{(n, 1)\}$.

On the left in the figures below, we present a matrix M with nonzero entries given by asterisks. On the right, asterisks denote the entries spanning the smallest subspace V_p containing M. Here, $p = \{(2, 1), (4, 2), (5, 3), (6, 5)\}$.

$$M = \begin{pmatrix} * & * & * & * & * & * \\ * & 0 & * & * & * & * \\ 0 & * & * & 0 & 0 & * \\ 0 & * & 0 & * & * & * \\ 0 & 0 & * & * & 0 & * \\ 0 & 0 & 0 & 0 & * & * \end{pmatrix}, \begin{pmatrix} * & * & * & * & * & * \\ * & * & * & * & * & * \\ 0 & * & * & * & * & * \\ 0 & * & * & * & * & * \\ 0 & 0 & * & * & * & * \\ 0 & 0 & 0 & 0 & * & * \end{pmatrix}.$$

Define $\mathcal{S}_p = V_p \cap \Sigma$. Thus, if V_p consists of Hessenberg matrices, $\mathcal{S}_p$ consists of symmetric, tridiagonal matrices (in particular, $\mathcal{S}_p$ contain the Jacobi matrices). Define also l_p^0 to be the set of strictly lower triangular matrices in V_p.

Representations of V_p, $\mathcal{S}_p$, l_p^0, p = $\{(2, 1), (4, 2), (5, 3), (6, 2)\}$, $n = 6$ are:

$$\begin{pmatrix} * & * & * & * & * & * \\ * & * & * & * & * & * \\ 0 & * & * & * & * & * \\ 0 & * & * & * & * & * \\ 0 & 0 & * & * & * & * \\ 0 & 0 & 0 & 0 & * & * \end{pmatrix}, \begin{pmatrix} * & * & 0 & 0 & 0 & 0 \\ * & * & * & * & 0 & 0 \\ 0 & * & * & * & * & 0 \\ 0 & * & * & * & * & 0 \\ 0 & 0 & * & * & * & * \\ 0 & 0 & 0 & 0 & * & * \end{pmatrix}, \begin{pmatrix} 0 & 0 & 0 & 0 & 0 & 0 \\ * & 0 & 0 & 0 & 0 & 0 \\ 0 & * & 0 & 0 & 0 & 0 \\ 0 & * & * & 0 & 0 & 0 \\ 0 & 0 & * & * & 0 & 0 \\ 0 & 0 & 0 & 0 & * & 0 \end{pmatrix}.$$

For a permutation $\pi \in S_N$ and a minimal set of entries p, set

$$\mathcal{Z}_p^\pi = \mathcal{Z}^\pi \cap V_p \simeq D^\pi \times l_p^0, \quad \mathcal{S}_p^\pi = \Sigma^\pi \cap V_p = \Sigma^\pi \cap \mathcal{S}_p.$$

Proposition 4.10 *The restriction $\phi_p^\pi : \mathcal{Z}_p^\pi \to \mathcal{S}_p^\pi$ is a diffeomorphism.*

Proof First, we verify $\phi_p^\pi(\mathcal{Z}_p^\pi) \subset \mathcal{S}_p^\pi$ and, for the restriction of $\psi_p^\pi = \phi_p^{\pi^{-1}}$, $(\psi_p^\pi)(\mathcal{S}_p^\pi) \subset \mathcal{Z}_p^\pi$. We prove the first claim: the second is similar. For $Z \in \mathcal{Z}_p^\pi$, write $Z = L^{-1} \Lambda^\pi L$ for $\Lambda^\pi \in D^\pi$, $L \in Lo^1$, and decompose $L = QR$. Then

$$RZR^{-1} = Q^T \Lambda^\pi Q = \phi^\pi(Z) \in \Sigma^\pi.$$

On the other hand, as $R \in Up$ and $Z \in \mathcal{Z}_p^\pi$, we have $RZR^{-1} \in V_p$. Thus, $\phi_p^\pi(Z) = \phi^\pi(Z) \in \Sigma^\pi \cap V_p = \mathcal{S}_p^\pi$. Now mimic the proof of Theorem 4.6. $\square$

We restrict again the maps $\phi_p^\pi : \mathcal{Z}_p^\pi \simeq D^\pi \times l_p^0 \to \mathcal{S}_p^\pi$ by fixing spectrum. For $\Lambda = \mathrm{diag}(\lambda_1 > \cdots > \lambda_N)$ and a profile p, define the *isospectral manifold* of profile p and spectrum equal to the spectrum of Λ,

$$\mathcal{O}_{\Lambda, p} = \{S \in V_p \cap \Sigma \mid \sigma(S) = \sigma(\Lambda)\}.$$

The set $\mathcal{O}_{\Lambda, p}$ is the intersection of the full isospectral manifold $\mathcal{O}_{\{(n,1)\}, \Lambda}$ with the subspace V_p. Standard arguments with group actions suffice to prove that $\mathcal{O}_{\{(n,1)\},\Lambda}$ is a manifold[1] for *arbitrary* real, diagonal matrices Λ, possibly with repeated eigenvalues. In this general case, not every subspace of Σ intercepts $\mathcal{O}_{\{(n,1)\},\Lambda}$ is a manifold: A simple exercise shows that real, symmetric 3×3 tridiagonal matrices with eigenvalues $\{0, 0, 1\}$ form a figure eight (two circles attached at a point). If Λ has simple eigenvalues, then the isospectral set of real, symmetric tridiagonal matrices is indeed a manifold (Tomei, 1984).

As in Theorem 4.8, Toda flows are defined by equation (4.1).

Theorem 4.11 *(charts) Fix $\Lambda^\pi \in D^\pi$ and p as above. The restriction*

$$\phi^\pi_{\Lambda,p} : l^0_p \to \mathcal{O}^\pi_{\Lambda,p} \equiv \phi^\pi_{\Lambda,p}(l^0_p), \quad Z^0 \mapsto \phi^\pi_p(Z^0 + \Lambda^\pi)$$

is a diffeomorphism onto the manifold $\mathcal{O}^\pi_{\Lambda,p}$. The sets $\mathcal{O}_{\Lambda,p}$ are connected, compact manifolds: the open dense sets $\mathcal{O}^\pi_{\Lambda,p}$ provide an atlas, $\mathcal{O}_{\Lambda,p} = \cup_{\pi \in S_N} \mathcal{O}^\pi_{\Lambda,p}$, with charts

$$\psi^\pi_{\Lambda,p} = \left(\phi^\pi_{\Lambda,p}\right)^{-1} : \mathcal{O}^\pi_{\Lambda,p} \to l^0_p.$$

Each $\mathcal{O}^\pi_{\Lambda,p}$ (and thus $\mathcal{O}_{\Lambda,p}$) is invariant under the Toda flows. In these charts, the Toda flow becomes the solution of an equation in l^0_p of the form

$$Z_{ii}(t) = Z_{ii}(0), \quad Z_{ij}(t) = c_{ij}(t) Z_{ij}(0).$$

All such flows commute. For Toda flows, $c_{ij}(t) = e^{t(f(\lambda_{\pi(i)}) - f(\lambda_{\pi(j)}))}$.

Proof Follow the proof of Lemma 4.3 to show that the sets $\mathcal{O}^\pi_{\Lambda,p}$ are open, dense sets covering $\mathcal{O}_{\Lambda,p}$. The evolution in Z-variables is given in Theorem 4.8. A simple computation proves that flows of the form above commute. □

Exercise 4.12 Show that $\mathcal{O}_{\Lambda,p}$ is a *submanifold* of Σ.

Corollary 4.13 *For entries $(i_0, j_0), (i, j)$ in a profile p for which $i_0 > j_0, i > j, (i_0, j_0) \neq (i, j)$, the functions*

$$I_{ij} = \frac{Z_{ij}}{Z_{i_0,j_0}} \frac{f(\lambda_{\pi(i_0)}) - f(\lambda_{\pi(j_0)})}{f(\lambda_{\pi(i)}) - f(\lambda_{\pi(j)})}$$

are integrals of the Toda flow. If $\mathcal{O}_{\Lambda,p}$ has dimension m, then the Toda flow on $\mathcal{O}_{\Lambda,p}$ has $m - 1$ integrals.

Thus the Toda flow is a so called *maximally super-integrable* system. For Jacobi real matrices, super-integrability was proved in Agrotis et al. (2006).

---

[1] All manifolds in this text are differentiable submanifolds of some matrix vector space.

4.2 Profiles and Isospectral Manifolds

Proof From Theorem 4.11,

$$\ln Z_{ij}(t) = \ln Z_{ij}(0) + t(f(\lambda_{\pi(i)}) - f(\lambda_{\pi(j)})).$$

Now simply differentiate the functions I_{ij} to show they are conserved quantities. To count dimensions, recall that $Z_{i,i}(t)$ is constant for all i. □

Remark 4.14 The methods in this section extend the purview of the Toda system in a very nontrivial way. The complete integrability of the Toda lattice in a variety of phase spaces was well established in the literature over the years. As noted following Corollary 4.13, we see that the Toda system is integrable, in fact maximally super-integrable, on a manifold with *arbitrary* profile V_p. In the symplectic/Lie-theoretic view of the Toda system, the natural restrictions of the full Toda system are to symplectic submanifolds such as tridiagonal matrices, or pentadiagonal matrices, and so on. Here $\mathcal{O}_{\Lambda,p}$ in general has no apparent symplectic or Lie-theoretic interpretation.

Remark 4.15 The construction of Z-variables, together with its properties related to the Toda lattice, carry through complex matrices (with real spectrum). A more general context accommodating both cases is provided in Torres and Tomei (2013). Thus, for example, Jacobi matrices, real and complex, are parametrized by bidiagonal matrices Z_π: The only difference is the fact that, in the real case, the offdiagonal entries z_j are strictly positive, while in the complex case, they are nonzero complex numbers. The results in Leite et al. (2023) provide the appropriate extension of Z-variables for nonsymmetric matrices with real, simple spectrum.

For completeness, we relate Moser's inverse variables for Jacobi matrices and Z-variables, as in Leite et al. (2008). Let $\mathcal{J}_\Lambda \subset \mathcal{O}_{\Lambda,p}$ be the Jacobi matrices with simple spectrum Λ. Parametrize the matrices in l_p^0 as

$$Z_\pi = \begin{pmatrix} \lambda_1^\pi & & & & \\ z_1^\pi & \lambda_2^\pi & & & \\ & z_2^\pi & \lambda_3^\pi & & \\ & & \ddots & \ddots & \\ & & & z_{n-1}^\pi & \lambda_N^\pi \end{pmatrix}.$$

Note that here $p = \{(2,1), (3,2), \ldots, (n, n-1)\}$. A simple computation yields

$Z_\pi = L^{-1} \Lambda^\pi L, L \in Lo^1$ for

$$L = \begin{pmatrix} 1 & 0 & 0 & \cdots & 0 \\ \frac{z_1^\pi}{\lambda_2^\pi - \lambda_1^\pi} & 1 & 0 & \cdots & 0 \\ \frac{z_1^\pi z_2^\pi}{(\lambda_3^\pi - \lambda_1^\pi)(\lambda_3^\pi - \lambda_2^\pi)} & \frac{z_2^\pi}{\lambda_3^\pi - \lambda_2^\pi} & 1 & \cdots & 0 \\ \vdots & & & & \vdots \\ \frac{z_1^\pi z_2^\pi \cdots z_{n-1}^\pi}{(\lambda_N^\pi - \lambda_1^\pi)(\lambda_N^\pi - \lambda_2^\pi)\cdots(\lambda_N^\pi - \lambda_{n-1}^\pi)} & \frac{z_2^\pi \cdots z_{n-1}^\pi}{(\lambda_N^\pi - \lambda_2^\pi)\cdots(\lambda_N^\pi - \lambda_{n-1}^\pi)} & \cdots & & 1 \end{pmatrix}.$$

Exercise 4.16 Compute L_π^{-1}.

Proposition 3.8 still holds if the eigenvalues of $X \in \mathcal{J}$ are ordered according to $\pi \in S_N$. Denote by $u_1^\pi, \ldots, u_n^\pi$ Moser's inverse variables for this ordering.

Proposition 4.17 *For any permutation π and any Jacobi matrix $J \in \mathcal{J}_\Lambda$,*

$$u_k^\pi = u_1^\pi \left| \frac{z_1^\pi \cdots z_{k-1}^\pi}{(\lambda_k^\pi - \lambda_1^\pi) \cdots (\lambda_k^\pi - \lambda_{k-1}^\pi)} \right| = u_1^\pi \left| \frac{\prod_{i=1}^{k-1} z_i^\pi}{\prod_{i=1}^{k-1}(\lambda_k^\pi - \lambda_i^\pi)} \right|, \quad k = 2, \ldots, n-1,$$

$$z_k^\pi = \left| \frac{\prod_{i=1}^{k}(\lambda_{k+1}^\pi - \lambda_i^\pi)}{\prod_{i=1}^{k-1}(\lambda_k^\pi - \lambda_i^\pi)} \right| \frac{u_{k+1}^\pi}{u_k^\pi}, k = 1, \ldots, n-1.$$

Proof From the factorization $L = QR$, the coordinates $\{u_i^\pi\}$ are obtained by normalizing the first column of the matrix L. □

From Theorem 4.11, we obtain substantial asymptotics information. Recall the projection $\pi_{s\ell}$ on strictly lower triangular matrices, defined in the proof of Theorem 4.8. Also, consider complementary projections π_s and π_{su}, taking values on symmetric and strictly upper triangular matrices. The nonlinearity $f : \mathbb{R} \to \mathbb{R}$ is *injective* if $f(\lambda_i) \neq f(\lambda_j), i \neq j$.

Given an equilibrium point E of a vector field in a manifold M, the *stable* and *unstable* sets E^s and E^u consist of the points $m \in M$, which are initial conditions of the associated flow satisfying respectively

$$\lim_{t \to \infty} m(t) = E, \quad \lim_{t \to -\infty} m(t) = E.$$

Theorem 4.18 (Asymptotics)) *For f injective, restrict the Toda flow to $\mathcal{O}_{\Lambda,p}$.*

(1) *The equilibria are the diagonal matrices Λ^π.*
(2) *Near Λ^π, $\phi_{\Lambda,p}^\pi : Z^0 \in L_p^0 \to X \in \mathcal{O}_{\Lambda,p}^\pi$ satisfies*

$$X = \Lambda^\pi + \pi_s Z^0 + O(\|Z^0\|^2), \quad Z^0 = \pi_{s\ell} X + O(\|X - \Lambda^\pi\|^2).$$

4.2 Profiles and Isospectral Manifolds 115

(3) *For an equilibrium $\Lambda^\pi \in \mathcal{O}^\pi_{\Lambda,p}$, the stable and unstable manifolds at $(\psi^\pi_{\Lambda,p})(\Lambda)$ are transversal vector spaces in $(\psi^\pi_{\Lambda,p})(\mathcal{O}^\pi_{\Lambda,p}) = L^0_p$, i.e., L^0_p is their direct sum.*

Proof From the explicit evolution of $Z(t)$ under the Toda flow in Theorem 4.8, it is clear that each chart domain $\mathcal{O}^\pi_{\Lambda,p}$ is invariant. To compute equilibria X, we consider Z-variables by showing that $Z(X)$ is diagonal. From Theorem 4.11, for $i \neq j$, $Z_{ij}(t) = e^{t(f(\lambda_{\pi(i)}) - f(\lambda_{\pi(j)}))} Z_{ij}(0)$. By hypothesis, at an equilibrium Z we must then have $Z_{ij} = 0$: Z is diagonal.

For the second item, recall the construction of $\phi^\pi_{\Lambda,p}$,

$$Z^0 \mapsto Z = Z^0 + \Lambda^\pi \to L \in Lo^1, \text{ from } Z = L^{-1}\Lambda^\pi L$$

$$\to Q \text{ orthogonal, from } L = QR \mapsto X = Q^T \Lambda^\pi Q.$$

In particular, we must compute standard derivatives at the points

$$Z_0 = 0 \to Z = \Lambda^\pi \to L = I \to Q = I \to X = \Lambda^\pi.$$

Consider $Z(t) = \Lambda^\pi + t\hat{Z}, \hat{Z} \in L^0_p$. We have $Z(t) = L^{-1}\Lambda^\pi L(t)$, where $L(t) = I + t\hat{L} + O(t^2)$ and $L(t) - I \in L^0_p$. Differentiating at $t = 0$, we find $\Lambda^\pi \hat{L} - \hat{L}\Lambda^\pi = \hat{Z}$, so that $\hat{L} \in L^0_p$, as λ^π has simple spectrum. Also $L(t) = Q(t)R(t)$, where $Q(t)$ is orthogonal and $R(t) \in Up^+$, $Q(t) = I + t\hat{Q} + O(t^2)$ and $R(t) = I + t\hat{R} + O(t^2)$. From $X(t) = Q^T(t)\Lambda^\pi Q(t)$, we find $\hat{X} = \Lambda^\pi \hat{Q} - \hat{Q}\Lambda^\pi$, as $\hat{Q}$ is necessarily skew symmetric. But $\hat{L} = \hat{Q} + \hat{R}$ and so, as $\hat{R} \in Up$, $\hat{Q} = \pi_k \hat{L}$. Thus

$$\hat{X} = \Lambda^\pi(\pi_k \hat{L}) - (\pi_k \hat{L})\Lambda^\pi = \Lambda^\pi \hat{L} - \hat{L}\Lambda^\pi - \Lambda^\pi(\pi_u \hat{L}) + (\pi_u \hat{L})\Lambda^\pi.$$

But $\pi_u \hat{L} = \hat{L}^T_- + \hat{L}_0 + \hat{L}_+ = \hat{L}_+$, and we see that $\hat{X} = \hat{Z} + \hat{Z}^T$ from the explicit form of the projections. This proves the first claim in the second item. On the other hand, from $\hat{X} = \hat{Z} + \hat{Z}^T$, we see directly that $\hat{Z} = \pi_{s\ell} X$.

For the third item, use $Z_{ij}(t) = e^{t(f(\lambda_{\pi(i)}) - f(\lambda_{\pi(j)}))} Z_{ij}(0)$ from Theorem 4.8: as $t \to \infty$, convergence to Λ^π occurs if and only if $Z_{ij}(0) = 0$ whenever $\lambda_{\pi(i)} > \lambda_{\pi(j)}$. A similar argument holds for $t \to -\infty$. □

As an application of charts, we rederive some results from Section 4.1. Let $e, \rho \in S_N$ be respectively the trivial and reversing permutations, i.e., $e(k) = k, \rho(k) = n + 1 - k$. Set

$$\Lambda = \Lambda^e = \text{diag}(\lambda_1 > \cdots > \lambda_N), \quad \Lambda^\rho = \text{diag}(\lambda_N < \cdots < \lambda_1).$$

Corollary 4.19 *Let $X(0) \in \Sigma^s$ and f injective.*

(1) *The orbit $X(t)$ converges to a diagonal matrix $X(\pm\infty)$, for $t \to \pm\infty$. If $X(0)$ is a Jacobi matrix, for $f(x) = x$, then $X(\infty) = \Lambda, X(-\infty) = \Lambda^\rho$.*

(2) If $X(\infty) \in \Sigma_\pi$, then, for $t \to \infty$ and $i \geq j$,
$$X_{ij}(t) = e^{t(f(\lambda_{\pi(i)}) - f(\lambda_{\pi(j)}))} Z_{ij}(0)(1 + \text{e.s.e.}).$$
The same formula holds if $X(-\infty) \in \Sigma_\pi$, for $t \to -\infty$.

The results provide asymptotic information for symmetric initial conditions with simple spectrum. The argument uses the fact that diagonal matrices belong to chart domains on which Z-variables (the images of charts) are well defined. This is not the case for Moser's inverse variables $\{b_i\}$, even for Jacobi matrices. For Jacobi matrices, as in the description of time reversal after (3.70), asymptotics as $t \to -\infty$ follow from Proposition 4.17 by changing between charts related to the trivial and reversing permutations, as shown below.

Proof We prove (1). As the set $\mathcal{O}_{\Lambda,p}$ is compact, take a subsequence of $(X(N))$, $n \in \mathbb{N}$. We must show that $X(t) \to X(\infty)$, a diagonal matrix: we use the appropriate chart. From Lemma 4.3 (1), $X(\infty)$ belongs to some chart $\mathcal{O}^\pi_{\Lambda,p}$. As charts are open sets of $\mathcal{O}_{\Lambda,p}$, for some $n_0 \in \mathbb{N}$ we have $X(n_0) \in \mathcal{O}^\pi_{\Lambda,p}$. From Theorem 4.11, charts are invariant under Toda flows, so that, for all $t \in \mathbb{R}$, $X(t) \in \mathcal{O}^\pi_{\Lambda,p}$. Again, from Theorem 4.11, $X(t)$ in Z-variables we have
$$Z_{ii}(t) = Z_{ii}(0) = Z_{ii}(\infty), \quad Z_{ij}(t) = e^{t(f(\lambda_{\pi(i)}) - f(\lambda_{\pi(j)}))} Z_{ij}(0) \to Z_{ij}(\infty),$$
and convergence as $t \to \infty$ holds if and only if $Z_{ij}(0) = 0$ whenever $f(\lambda_{\pi(i)}) > f(\lambda_{\pi(j)})$. In this case, clearly, $Z(t) \to \Lambda^\pi$ and the limit X_∞ then is a diagonal matrix. The argument for $t \to -\infty$ is similar, after change of charts: the permutation inducing the appropriate chart is defined by $\Lambda^\pi = X(-\infty)$.

From Lemma 4.3 (4), if $X(0)$ is a Jacobi matrix, then, for all permutations π, $X(t)$ is π-admissible, so that $X(0) \in \mathcal{O}^\pi_{\Lambda,p}$ for all $\pi \in S_N$. From Theorem 4.18, $X(t) \in \mathcal{O}^\pi_{\Lambda,p}$. For the limit $t \to \infty$, consider $\pi = e$, the trivial permutation. Then $\Lambda \in \mathcal{O}^\pi_{\Lambda,p}$ and the formulas for $Z(t)$ give $Z(\infty) = X(\infty) = \Lambda$. Similarly, for $\mathcal{O}^\pi_{\Lambda,p}$, we obtain $X(-\infty) = \Lambda_\rho$.

For (2) with $t \to \infty$, by Theorem 4.18 (2), $X(t) = \phi^\pi_{\Lambda,p}(Z(t))$,
$$X(t) = \phi^\pi_{\Lambda,p}(Z(t)) = \Lambda^\pi + \pi_s Z(t) + O(\|Z(t)\|^2),$$
near $\Lambda^\pi = X(\infty)$. But again, for $i \geq j$,
$$Z_{ij}(t) = e^{t(f(\lambda_{\pi(i)}) - f(\lambda_{\pi(j)}))} Z_{ij}(0).$$
The case $t \to -\infty$ is similar. □

4.2 Profiles and Isospectral Manifolds

We present an alternative computation of the phase shifts (3.82) of the standard Toda flow, following the argument in Leite et al. (2008). The idea is to express the long-term behavior of the physical variables $\{x_k(t), y_k(t)\}$ in terms of Z-variables $\{z_k^\pi(t)\}$. Keeping the notation of that section, we assume that J is an $N \times N$ Jacobi matrix, and, by Corollary 4.19, $\lim_{t\to\infty} J(t) = \Lambda$, so that $\pi = e$ induces a chart centered at $J(\infty)$.

First, write equation (3.72)

$$x_k(t) - x_{k+1}(t) = 2\ln(2b_k(t))$$

in Z-variables: from Lemma 4.3 (4), Theorem 4.18 (2), and Corollary 4.19,

$$b_k(t) = z_k(t) + O(\|Z(t) - \Lambda\|^2) = e^{t(\lambda_{k+1}-\lambda_k)} z_k(0)(1 + \text{e.s.e.}),$$

provided $z_k(0) \neq 0$, so that

$$x_k - x_{k+1} = 2t(\lambda_{k+1} - \lambda_k) + 2\ln(2z_k(0)) + \text{e.s.e.}$$

Exercise 4.20 Rederive equation (3.73) by using Proposition 4.17.

Again, as linear momentum is conserved, equation (3.75) holds,

$$\sum_{k=1}^N x_k(t) = \sum_{k=1}^N x_k(0) - 2(\operatorname{tr}\Lambda) t .$$

From the differences $x_k - x_{k+1}$, as in the computations after (3.75) one obtains the differences $x_1 - x_k$. Adding and using the formulae above,

$$x_N(t) = -2\lambda_N t + \frac{1}{n}\sum_{i=1}^N x_i(0) - \frac{2}{N}\sum_{i=1}^{N-1} i \ln(2z_i(0)) + \text{e.s.e.}$$

from which, exploiting the consecutive differences $x_k - x_{k+1}$, we obtain the counterpart to equation (3.77),

$$x_k(t) = -2\lambda_k t + \frac{1}{N}\sum_{i=1}^N x_i(0) - \frac{2}{N}\sum_{i=1}^{N-1} i \ln(2z_i(0)) + 2\sum_{j=k}^{N-1} \ln(2z_j(0)) + \text{e.s.e.}$$

$$= -2\lambda_k t + \beta_k^+ + \text{e.s.e.}, \quad t \to \infty,$$

the asymptotic behavior of the N particles under the standard Toda flow in terms of $Z = \psi_{\Lambda,p}^e(J)$. Notice that the β_k^+ depend on the Z-variables $z_j(0)$: In equation (3.72), the b_k depends instead on Moser's variables $u_i(1)$.

By performing time reversal (i.e., taking the chart associated with $\pi = \rho$, the reversing permutation), as in Corollary 4.19, we obtain

$$x_{\rho(k)}(t) = -2\lambda_k t + \beta_k^- + \text{e.s.e.} \quad \text{for } t \to -\infty .$$

and again

$$x_{\rho(k)}(t) - x_{\rho(k+1)}(t) = 2t(\lambda_{k+1} - \lambda_k) + 2\ln 2z_k^\rho$$

from which we derive

$$x_{\rho(k)}(t) = -2\lambda_k t + \frac{1}{N}\sum_{i=1}^{N} x_i(0) - \frac{2}{N}\sum_{i=1}^{N-1} i\ln(2z_i^\rho(0)) + 2\sum_{j=k}^{N-1} \ln(2z_j^\rho(0)) + \text{e.s.e.}$$

$$= -2\lambda_k t + \beta_{\rho(k)}^- + \text{e.s.e.}, \quad t \to \infty.$$

We now relate the Z-coordinates $\{z_i\}$ and $\{z_i^\rho\}$ in charts associated with the trivial and the reversing permutation by using Proposition 4.17:

$$z_k = \left|\frac{\prod_{i=1}^{k}(\lambda_{k+1} - \lambda_i)}{\prod_{i=1}^{k-1}(\lambda_k - \lambda_i)}\right| \frac{u_{k+1}}{u_k},$$

$$z_{N-k}^\rho = \left|\frac{\prod_{i=1}^{N-k}(\lambda_{N-k+1}^\rho - \lambda_i^\rho)}{\prod_{i=1}^{N-k-1}(\lambda_{N-k}^\rho - \lambda_i^\rho)}\right| \frac{u_{N-k+1}^\rho}{u_{N-k}^\rho} = \left|\frac{\prod_{i=1}^{N-k}(\lambda_k - \lambda_{N-i+1})}{\prod_{i=1}^{N-k-1}(\lambda_{k+1} - \lambda_{N-i+1})}\right| \frac{u_k}{u_{k+1}},$$

so that $\ln z_k = -\ln x_{N-k}^\rho + \delta_k$, where δ is a combination of terms of the form $\ln(\lambda_i - \lambda_j)$. The computation of the shift $\beta_k^+ - \beta_{\rho(k)}^-$ now is immediate.

Exercise 4.21 Prove the first statement of Corollary 4.19 for arbitrary symmetric initial conditions, by taking limits.

Shub and Vasquez (1987) used the *LU* decomposition in their proof that the Toda flow and the QR algorithm (as a diffeomorphism) on the isospectral manifold of full, symmetric matrices are *Morse–Smale* systems, a property which implies *structural stability*. Leite et al. (2008, 2010, 2013) considered Z-variables on tridiagonal matrices, leading to a detailed description of a fundamental algorithm to compute eigenvalues, the inverse iteration with Wilkinson shifts. Torres and Tomei (2013) extended the constructions to arbitrary profiles and non-compact real semi-simple Lie algebras. In Leite et al. (2023), the authors extend Z-variables to nonsymmetric matrices with real spectrum and show that the properties related to linearization of the Toda flows, super-integrability, and profile preservation still hold.

4.3 The Tridiagonal Isospectral Manifold

We now consider real tridiagonal, symmetric matrices, and isospectral manifolds within, for which $p = \{(2,1), (3,2), \ldots, (n, n-1)\}$. Jacobi matrices

4.3 The Tridiagonal Isospectral Manifold

$X \in \mathcal{J}$ are special symmetric tridiagonal matrices: subdiagonal entries $(2, 1)$, ..., $(n, n-1)$ are strictly positive. A tridiagonal matrix is *nondegenerate* if its subdiagonal entries are nonzero. Jacobi matrices have simple spectrum, but this is not necessarily the case for tridiagonal matrices.

From Lemma 4.3, X is π-admissible for all $\pi \in S_N$, so that $X \in \Sigma^\pi$. We consider an example, from which the general situation will be clear. Let $\Lambda = \text{diag}(7, 5, 4)$ and a permutation $\pi \in S_3$ for which $\Lambda^\pi = \text{diag}(7, 4, 5)$. The interior of the polygon in Figure 4.1 is $\mathcal{O}^\pi_{\Lambda, p}$.

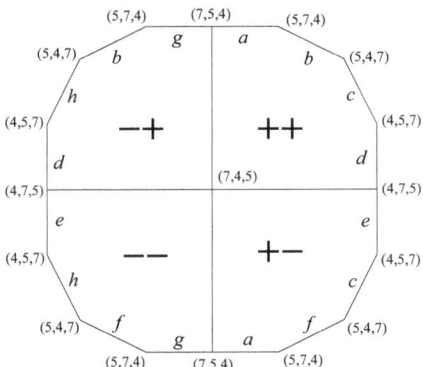

Figure 4.1 $\mathcal{O}^\pi_{\Lambda, p}$, for $\Lambda^\pi = \text{diag}(7, 4, 5)$.

There are four connected sets of nondegenerate matrices: Each set is associated to a choice of signs for the two subdiagonal entries. More precisely, each such set is of the form $E \mathcal{J} E$ for some diagonal matrix E with ± 1s along the diagonal (a sign diagonal matrix). At points on the cross in Figure 4.1 correspond to points where the usual variables break down: Some eigenvector has a nonzero first coordinate. The associated inverse algorithm still works for the horizontal edges, but not at vertical edges. Still, the edges in the cross lie in $\mathcal{O}^\pi_{\Lambda, p}$, where the Z_{ij} are good variables: local computations are easy to perform.

Changes of sign mix well with Z-variables. As before, let $\psi^\pi = (\phi^\pi)^{-1} : \Sigma^\pi \to \mathcal{Z}^\pi$ take X to $Z = Z(X)$.

Proposition 4.22 *For a diagonal sign matrix E,*

$$\psi^\pi(EXE) = E \psi^\pi(X) E, \text{ or equivalently, } \phi^\pi(EZE) = E \phi^\pi(Z) E.$$

Thus, once we know the behavior of the Toda flow in one quadrant, we know the behavior in the other quadrants.

Proof We prove the second statement, by computing $\phi^\pi(Z)$ and $\phi^\pi(EZE)$ side by side. For $Z = L^{-1}\Lambda^\pi L$, in order to keep the diagonal entries of L equal to 1 and the diagonal entries of R positive, $EZE = EL^{-1}E\Lambda^\pi ELE$. Now $L = QR$ induces $ELE = EQE\,ERE$. By definition,

$$\phi^\pi(EZE) = (EQE)^T \Lambda^\pi EQE = E\phi^\pi(Z)E\,. \qquad \square$$

The vertices of the four hexagons consist of diagonal matrices, with eigenvalues along the diagonal as indicated. At the center lies Λ^π. Segments joining vertices consist of degenerate matrices, which have a zero either in entry $(1, 2)$ (as the matrices represented by a segment joining $(7, 4, 5)$ and $(7, 5, 4)$) or in entry $(2, 3)$ (as along the segment joining $(7, 4, 5)$ and $(4, 7, 5)$). The interior of the hexagon labeled ++ consists of Jacobi matrices: From Lemma 3.5, it is homeomorphic to the interior of the positive orthant in $\mathbb{R}^3$.

To obtain the isospectral manifold $\mathcal{O}_{\Lambda,p}$, glue boundary segments with the same label. The reader is invited to verify that the resulting set is a bitorus. Clearly, $\mathcal{O}^\pi_{\Lambda,p}$ is a dense, open set of $\mathcal{O}_{\Lambda,p}$.

The interior of each hexagon is invariant under the Toda flow, as is every edge and vertex. By Theorem 3.11, as $t \to \pm\infty$ the orbit approaches a vertex. From Corollary 4.19, initial conditions in the interior of the hexagons go to Λ or $\Lambda_\rho = \text{diag}(4, 5, 7)$ as $t \to \infty$ or $-\infty$ respectively.

Example 4.23 If S is a real symmetric 2×2 matrix with eigenvalues $\Lambda = \text{diag}(\lambda_1, \lambda)$, $\lambda_1 > \lambda_2$, then $S = S(\theta) = Q^T(\theta)\Lambda Q(\theta)$, where

$$Q^T(\theta) = \begin{pmatrix} \cos\theta & \sin\theta \\ \sin\theta & -\cos\theta \end{pmatrix}, \quad -\pi/2 \le \theta < \pi/2\,.$$

Also, for $p = \{(2, 1)\}$, $\mathcal{O}^e_{\Lambda,p} = \{S = S(\theta) : -\pi/2 < \theta < \pi/2\}$, $\mathcal{O}_{\Lambda,p} = \{S = S(\theta) \, (\theta \mod \pi)\}$. Note that $\mathcal{O}^e_{\Lambda,p}$ is dense in $\mathcal{O}_{\Lambda,p}$, which is a manifold (a circle!), as $S(\pi/2) = \text{diag}(\lambda_2, \lambda_1) = S(-\pi/2)$. Moreover, $\text{diag}(\lambda_2, \lambda_1) \in \mathcal{O}_{\Lambda,p} \setminus \mathcal{O}^e_{\Lambda,p}$, by item (1) of Lemma 4.3. However, $\text{diag}(\lambda_2, \lambda_1) \in \mathcal{O}^\pi_{\Lambda,p}$ where π interchanges λ_1 and λ_2. Clearly, $\mathcal{O}_{\Lambda,p} = \mathcal{O}^e_{\Lambda,p} \cup \mathcal{O}^\pi_{\Lambda,p}$, consistent with Lemma 4.2. Jacobi matrices $S(\theta)$ correspond to $0 < \theta < \pi/2$, whereas $S(\theta)$ has negative off-diagonal entries for $-\pi/2 < \theta < 0$. Also, for initial conditions $S(\theta)$, $\theta \ne 0$ or $\pi/2 \mod \pi$,

$$S(\theta) \to \begin{pmatrix} \lambda_1 & 0 \\ 0 & \lambda_2 \end{pmatrix} \text{ as } t \to \infty \text{ and } S(\theta) \to \begin{pmatrix} \lambda_2 & 0 \\ 0 & \lambda_1 \end{pmatrix} \text{ as } t \to -\infty\,.$$

Also note that for $0 < \theta < \pi/2$, the first components u_1, u_2 of the eigenvectors corresponding to λ_1, λ_2 respectively, are given by the first row of $Q^T(\theta)$, $u_1 = \cos\theta > 0$, $u_2 = \sin\theta > 0$. By (3.166), the action-angle variables are

4.3 The Tridiagonal Isospectral Manifold

λ_1 and $\theta_1 = \ln\left(\frac{u_1}{u_2}\left|\frac{p'(\lambda_1)}{p'(\lambda_2)}\right|^{1/2}\right)$, $\{\theta_1, \lambda_1\} = 1$, where $p(\lambda) = \det(\lambda - S(\theta))$. We have $\left|\frac{p'(\lambda_1)}{p'(\lambda_2)}\right|^{1/2} = \left|\frac{\lambda_1 - \lambda_2}{p'(\lambda_2 - \lambda_1)}\right|^{1/2} = 1$, so $\theta_1 = \ln u_1/u_2$, or $u_1/u_2 = e^{\theta_1}$.

Using $u_1^2 + u_2^2 = 1$, we find $u_1 = e^{\theta_1}/\sqrt{1 + e^{2\theta_1}}$, $u_2 = 1/\sqrt{1 + e^{2\theta_1}}$. Inserting the formulae into $S(\theta) = Q^T(\theta)\Lambda Q(\theta)$, we obtain

$$S(\theta) = \frac{1}{1 + e^{2\theta_1}}\begin{pmatrix} \lambda_1 e^{2\theta_1} + \lambda_2 & (\lambda_1 - \lambda_2)e^{\theta_1} \\ (\lambda_1 - \lambda_2)e^{\theta_1} & \lambda_2 + \lambda_1 e^{2\theta_1} \end{pmatrix}. \tag{4.5}$$

Note that $Q(\theta)$ does not have an LU-positive decomposition. Indeed, for a Jacobi matrix $S = Q^T \Lambda Q$, $\Lambda = \text{diag}(\lambda_1, \lambda_2)$, $\lambda_1 > \lambda_2$, Proposition 3.5 is not consistent with the requirement that Q has such a decomposition, in the sense that if the first components of the normalized eigenvectors, i.e., the entries in the first row of Q^T, are chosen to be positive, then Q does not have an LU-positive decomposition. As in Lemma 4.2 (2), $Q_1 = EQ$ with $E = \text{diag}(1, -1)$, does have an LU-positive decomposition and again $S = Q_1^T \Lambda Q_1$, but now the first components of the normalized eigenvectors are no longer all positive.

Defining $Q_1(\theta) := EQ(\theta)$ as above, we again have $S(\theta) = Q_1^T \Lambda Q_1$, but now $Q_1(\theta)$ has an LU- positive decomposition. We find

$$L = \begin{pmatrix} 1 & 0 \\ \ell & 1 \end{pmatrix},$$

where $\ell = -\sin\theta/\cos\theta = -e^{-\theta_1}$.

Thus,

$$S(\theta) = \frac{1}{1 + \ell^2}\begin{pmatrix} \lambda_1 + \lambda_2 \ell^2 & (\lambda_2 - \lambda_1)\ell \\ (\lambda_2 - \lambda_1)\ell & \lambda_2 + \lambda_1 \ell^2 \end{pmatrix}. \tag{4.6}$$

Near the point $S(\theta) \sim \Lambda = \text{diag}(\lambda_1, \lambda_2)$, $\theta \sim 0$, the advantage of the Z variables over the action-angle variables is clear from 4.5 and 4.6. Whereas $\ell = -\sin\theta/\cos\theta$ covers smoothly through $\theta = 0$ on $\mathcal{O}^e_{\Lambda,p}$, the action-angle variables breaks down and must flip, $e^{\theta_1} \to -e^{\theta_1}$.

The isospectral tridiagonal manifold is an interesting topological object, first considered in Tomei (1984). Its universal covering is $\mathbb{R}^{N-1}$, but it does not admit a hyperbolic structure. Its Betti numbers are familiar combinatorial numbers. Fried (1986) proved that its cohomology ring is free. Gaifullin (2009) showed that finite covers of the manifold answer a question raised by Steenrod about general homology classes.

5
Random Matrix Ensembles

5.1 Invariant and Wigner Ensembles

The following definitions are taken from Bourgade and Yau (2017), Erdős et al. (2013), and Deift (2000); the first condition appeared initially in Erdős et al. (2012) and was made more explicit in Erdős et al. (2013). These are the two classes of random matrices to which our results apply.

Definition 5.1 (Generalized Wigner Ensemble (WE)) A generalized Wigner matrix is a real symmetric ($\beta = 1$) or Hermitian ($\beta = 2$) matrix $H = (H_{ij})_{i,j=1}^{N}$ such that H_{ij} are independent random variables for $i \leq j$ with

$$\mathbb{E} H_{ij} = 0, \quad \sigma_{ij}^2 := \mathbb{E}|H_{ij}|^2.$$

Next, assume there is a fixed constant $\upsilon > 0$ (independent of N, i, j) such that

$$\mathbb{P}(|H_{ij}| > x\sigma_{ij}) \leq \upsilon^{-1} \exp(-x^{\upsilon}), \quad x > 0.$$

Finally, assume there exist $C_1, C_2 > 0$ such that for all i, j,

$$\sum_{i=1}^{N} \sigma_{ij}^2 = 1, \quad \frac{C_1}{N} \leq \sigma_{ij}^2 \leq \frac{C_2}{N},$$

and for $\beta = 2$, for $i \neq j$, the matrix

$$\Sigma_{ij} = \begin{pmatrix} \mathbb{E}(\mathrm{Re}\, H_{ij})^2 & \mathbb{E}(\mathrm{Re}\, H_{ij})(\mathrm{Im}\, H_{ij}) \\ \mathbb{E}(\mathrm{Re}\, H_{ij})(\mathrm{Im}\, H_{ij}) & \mathbb{E}(\mathrm{Im}\, H_{ij})^2 \end{pmatrix}$$

has its smallest eigenvalue $\lambda_{\min}$ satisfying $\lambda_{\min} \geq C_1 N^{-1}$.

Definition 5.2 (Invariant Ensemble (IE)) Let $V : \mathbb{R} \to \mathbb{R}$ satisfy $V \in C^4(\mathbb{R})$, $\inf_{x \in \mathbb{R}} V''(x) > 0$ and $V(x) > (2 + \delta)\log(1 + |x|)$ for sufficiently large x and some fixed $\delta > 0$. Then we define an invariant ensemble[1] to be the set of all $N \times N$ symmetric ($\beta = 1$) or Hermitian ($\beta = 2$) matrices $H = (H_{ij})_{i,j=1}^{N}$ with probability density

$$\frac{1}{Z_N} e^{-N\frac{\beta}{2}\mathrm{tr} V(H)} \mathrm{d}H.$$

Here, $\mathrm{d}H = \prod_{i \leq j} \mathrm{d}H_{ij}$ if $\beta = 1$ and $\mathrm{d}H = \prod_{i=1}^{N} \mathrm{d}H_{ii} \prod_{i<j} \mathrm{d}\,\mathrm{Re}\, H_{ij} \mathrm{d}\,\mathrm{Im}\, H_{ij}$ if $\beta = 2$.

5.2 Estimates from Random Matrix Theory

We now introduce the results from random matrix theory that are needed to prove Theorem 6.2 and Proposition 6.5 in Chapter 6. Let H be an $N \times N$ Hermitian (or real symmetric) matrix with eigenvalues $\lambda_1 \geq \lambda_2 \geq \cdots \geq \lambda_N$ and let $\beta_1, \beta_2, \ldots, \beta_N$ denote the absolute value of the first components of the normalized eigenvectors. We assume the entries of H are distributed according to an invariant or generalized Wigner ensemble (see Section 5.1). Define the averaged empirical spectral measure

$$\mu_N(z) = \mathbb{E}\frac{1}{N} \sum_{i=1}^{N} \delta(\lambda_i - z),$$

where the expectation is taken with respect to the given ensemble, using the Riesz–Markov–Kakutani representation theorem.

Theorem 5.3 (Equilibrium measure, Bourgade et al., 2014) *For any WE or IE the measure μ_N converges weakly to a measure μ, called the equilibrium measure, which has support on a single interval $[a_V, b_V]$ and, for suitable constants C_μ and c_V, has a density ρ that satisfies*

$$\rho(x) \leq C_\mu \sqrt{b_V - x} \chi_{(-\infty, b_V]}(x) \tag{5.1}$$

and

$$\rho(x) = \frac{c_V}{\pi} \sqrt{b_V - x}(1 + O(b_V - x)) \tag{5.2}$$

as $x \uparrow b_V$.

Remark 5.4 For potentials $V(x)$ more general than those in Definition 5.2, the equilibrium measure may be supported on more than one interval, and

---

[1] This is not the most general class of V that one may consider but for technical reasons, as explained in Remark 1.4, we restrict our analysis to convex V.

the behavior near the edge can be of the form $\sim (b_V - x)^{\frac{1}{2}+2k}$, $k = 0, 1, 2, \ldots$ (see, e.g., Deift et al., 1999 and the references therein).

With the chosen normalization for WEs, the parameters are:

$$\sum_{i=1}^{N} \sigma_{ij}^2 = 1, \quad [a_V, b_V] = [-2, 2] \quad \text{and} \quad c_V = 1.$$

One can vary the support as desired by shifting and scaling, $H \to aH + bI$: the constant c_V then changes accordingly. When the entries of H are distributed according to a WE or an IE, with high probability (see Theorem 5.9) the top three eigenvalues are distinct and $\beta_j \neq 0$ for $j = N, N-1, N-2$. Next, let $d\mu$ denote the limiting spectral density, i.e., the equilibrium measure for the ensemble as $N \to \infty$. Then define γ_n to be the smallest real number such that

$$\frac{n-1}{N} = \int_{\gamma_n}^{\infty} d\mu.$$

Thus $\{\gamma_n\}$ represent the quantiles of the equilibrium measure.

There are four fundamental parameters involved in our calculations:

- first, we fix $0 < \sigma < 1$ once and for all;
- then we fix $0 < p < \sigma/4 < 1/3$;
- then we choose $s < \min\{\sigma/44, p/8\}$; and
- then, finally, $0 < c \leq 10/\sigma$ will be a constant that will allow us to estimate the size of various sums.

The specific meanings of the first three parameters is given below. Also, C denotes a generic constant that can depend on σ or p but not on s or N. We also make statements that will be "true for N sufficiently large." This should be taken to mean that there exists $N^* = N^*(\mu, \sigma, s, p)$ such that the statement is true for $N > N^*$. For convenience, in what follows we use the notation $\epsilon = N^{-\alpha/2}$.

Remark 5.5 Looking ahead to Chapter 6, we note that (ϵ, N) are in the scaling region in the sense of Definition 6.1 if and only if $\alpha - 10/3 \geq \sigma > 0$, and $\alpha = \alpha_N$ is allowed to vary with N.

Conditions 5.1 and 5.2 that follow play a key role in the proof of Theorem 6.2 and Proposition 6.5. These universality results provide the basis motivation for this monograph. Note that Conditions 5.1 and 5.2 hold with arbitrarily high probability as described in Theorems 5.9 and 5.10.

5.2 Estimates from Random Matrix Theory

Condition 5.1 *For* $0 < p < \sigma/4$, $\lambda_2 - \lambda_3 \geq p(\lambda_1 - \lambda_2)$.

Let $G_{N,p}$ denote the set of matrices that satisfy this condition.

Condition 5.2 *For any fixed* $0 < s < \min\{\sigma/44, p/8\}$,

(1) $\beta_n \leq N^{-1/2+s/2}$ *for all* n,
(2) $N^{-1/2-s/2} \leq \beta_n$ *for* $n = 1, 2$,
(3) $N^{-2/3-s} \leq \lambda_1 - \lambda_{n+1} \leq N^{-2/3+s}$, *for* $n = 1, 2$, *and*
(4) $|\lambda_n - \gamma_n| \leq N^{-2/3+s}(\min\{n, N - n + 1\})^{-1/3}$ *for all* n.

Let $R_{N,s}$ denote the set of matrices that satisfy these conditions (1)–(4).

Condition (4) is often referred to as the rigidity of the eigenvalues.

Remark 5.6 It is known that the distribution of the eigenvectors for IEs (namely, the Haar measure on the unitary or orthogonal group) depends only on $\beta = 1, 2$. And, if $V(x) = x^2$, the IE is also a WE. Therefore, if one can prove a general statement about the eigenvectors for WEs, then it must also hold for IEs. Stronger results can be proved for the eigenvectors for IEs, see Stam (1982) and Jiang (2006) for example.

The following theorem has its roots in the pursuit of the proof of universality in random matrix theory. See Tracy and Widom (1994) for the seminal result when $V(x) = x^2$ and $\beta = 2$. Further extensions include the works of Soshnikov (1999) and Tao and Vu (2010) for Wigner ensembles and Deift and Gioev (2007) and Shcherbina (2009) for invariant ensembles.

Theorem 5.7 *For both IEs and WEs, the variables $N^{1/2}(\beta_1, \beta_2, \beta_3)$ converge jointly in distribution to $(|X_1|, |X_2|, |X_3|)$, where $\{X_1, X_2, X_3\}$ are independent and identically distributed (i.i.d.) real ($\beta = 1$) or complex ($\beta = 2$) standard normal random variables. Additionally, for IEs and WEs,*

$$c_V^{2/3} 2^{-2/3} N^{2/3}(b_V - \lambda_1, b_V - \lambda_2, b_V - \lambda_3)$$

converges jointly in distribution to random variables $(\Lambda_{1,\beta}, \Lambda_{2,\beta}, \Lambda_{3,\beta})$ which are the smallest three eigenvalues $(\Lambda_{j,\beta} \leq \Lambda_{j+1,\beta})$ of the so-called stochastic Airy operator. Furthermore, $(\Lambda_{1,\beta}, \Lambda_{2,\beta}, \Lambda_{3,\beta})$ are distinct with probability one.

Proof The first claim follows from Bourgade and Yau (2017, Theorem 1.2). The second claim follows from Bourgade et al. (2014, Corollary 2.2 & Theorem 2.7). The last claim follows from Ramírez et al. (2011, Theorem 1.1). □

Definition 5.8 The distribution function $F_\beta^{\mathrm{gap}}(t)$ for $\beta = 1, 2$ and $t > 0$ is given by

$$F_\beta^{\mathrm{gap}}(t) = \mathbb{P}\left(\frac{1}{\Lambda_{2,\beta} - \Lambda_{1,\beta}} \leq t\right) = \lim_{N \to \infty} \mathbb{P}\left(\frac{1}{c_V^{2/3} 2^{-2/3} N^{2/3}(\lambda_1 - \lambda_2)} \leq t\right).$$

Properties of $G_\beta(t) := 1 - F_\beta^{\mathrm{gap}}(1/t)$, the distribution function for the first gap, are examined in Perret and Schehr (2014), Witte et al. (2013), and Monthus and Garel (2013) including the behavior of $G_\beta(t)$ near $t = 0$, which is critical for understanding which moments of $F_\beta'(t)$ exist.

The remaining theorems in this section are compiled from results that have been obtained recently in the literature. These results show that the conditions described above hold with arbitrarily high probability.

Theorem 5.9 *For WEs or IEs Condition 5.2 holds with high probability as $N \to \infty$, i.e., for any $s > 0$,*

$$\mathbb{P}(R_{N,s}) = 1 + o(1).$$

Proof We first consider WEs. The fact that Condition 5.2.(1) is satisfied with high probability follows from Erdős et al. (2012, Theorem 2.1) using estimates on the $(1, 1)$ entry of the Green's function. See Erdős (2012, Section 2.1) for a discussion of the use of these estimates. The fact that Condition 5.2.(2) and (3) are satisfied with high probability follows from Theorem 5.7 using Corollary 6.18. And finally, the statement that Condition 5.2.(4) is satisfied with high probability as $N \to \infty$ is a statement of the rigidity of eigenvalues, the main result of Erdős et al. (2012). Following Remark 5.6, we then have that Condition 5.2.(1) and (2) are satisfied with high probability for IEs.

For IEs, the fact that Condition 5.2.(4) is satisfied with high probability follows from Bourgade and Yau (2017, Theorem 2.4). And again, the fact that Condition 5.2.(3) is satisfied with high probability follows from Theorem 5.7 using Corollary 6.17. □

Theorem 5.10 *For both WEs and IEs,*

$$\lim_{p \downarrow 0} \limsup_{N \to \infty} \mathbb{P}(G_{N,p}^c) = 0,$$

where ^{c} denotes the complement.

Proof It follows from Theorem 5.7 that

$$\limsup_{N \to \infty} \mathbb{P}(G_{N,p}^c) = \lim_{N \to \infty} \mathbb{P}(\lambda_2 - \lambda_3 < p(\lambda_1 - \lambda_2))$$

$$= \mathbb{P}(\Lambda_{3,\beta} - \Lambda_{2,\beta} < p(\Lambda_{2,\beta} - \Lambda_{1,\beta})).$$

Then
$$\lim_{p \downarrow 0} \mathbb{P}(\Lambda_{3,\beta} - \Lambda_{2,\beta} < p(\Lambda_{2,\beta} - \Lambda_{1,\beta}))$$
$$= \mathbb{P}\left(\bigcap_{p>0} \{\Lambda_{3,\beta} - \Lambda_{2,\beta} < p(\Lambda_{2,\beta} - \Lambda_{1,\beta})\}\right)$$
$$= \mathbb{P}(\Lambda_{3,\beta} = \Lambda_{2,\beta}).$$

But from Ramírez et al. (2011, Theorem 1.1), $\mathbb{P}(\Lambda_{3,\beta} = \Lambda_{2,\beta}) = 0$. □

Throughout what follows, we assume that we are given a WE or an IE.

5.3 Technical Lemmas

Define $\delta_j = 2(\lambda_1 - \lambda_j)$ and $I_c = \{2 \leq n \leq N : \delta_n/\delta_2 \geq 1 + c\}$ for $c > 0$.

Lemma 5.11 *Let $0 < c < 10/\sigma$. Given Condition 5.2,*
$$|I_c^c| \leq N^{2s}$$
for N sufficiently large, where here the superscript c denotes the complement relative to $\{2, \ldots, N\}$.

Proof We use rigidity of the eigenvalues, Condition 5.2.(4). So,
$$|\lambda_n - \gamma_n| \leq N^{-2/3+s}(\hat{n})^{-1/3},$$
where $\hat{n} = \min\{n, N - n + 1\}$. Recall
$$I_c^c \subset \{2 \leq n \leq N : \lambda_1 - \lambda_n < (1 + c)(\lambda_1 - \lambda_2)\}.$$
Define
$$J_c = \{2 \leq n \leq N : \gamma_1 - \gamma_n \leq (2 + c + (\hat{n})^{-1/3})N^{-2/3+s}\}.$$
If $n \in I_c^c$ then
$$\lambda_1 - \lambda_n \leq (1 + c)N^{-2/3+s},$$
$$\gamma_1 - N^{-2/3+s} - (\gamma_n + (\hat{n})^{-1/3}N^{-2/3+s}) \leq \lambda_1 - \lambda_n \leq (1 + c)N^{-2/3+s},$$
$$\gamma_1 - \gamma_n \leq (2 + c + (\hat{n})^{-1/3})N^{-2/3+s},$$
and hence $n \in J_c$. Then to compute the asymptotic size of the set J_c let n^* be the largest element of J_c. Then $|J_c| = n^* - 1$ so that
$$\frac{n^* - 1}{N} = \int_{\gamma_{n^*}}^{\infty} d\mu, \quad |I_c^c| \leq |J_c| = n^* - 1 = N \int_{\gamma_{n^*}}^{\infty} d\mu.$$

Then using the estimates in Definition 5.3,

$$|I_c^c| \leq N \int_{\gamma_{n*}}^{\infty} d\mu \leq C_\mu N \int_{\gamma_{n*}}^{b_V} \sqrt{b_V - x} \, dx = N \frac{2C_\mu}{3} (\gamma_1 - \gamma_{n*})^{3/2}$$

$$\leq \frac{2C_\mu}{3} (3+c)^{3/2} N^{3s/2}.$$

Then because σ is fixed and hence c has an upper bound, and $s > 0$, $|I_c^c| \leq N^{2s}$ for sufficiently large N. □

We use the notation $\nu_n = \beta_n^2/\beta_1^2$ and note that for a matrix in $R_{N,s}$ we have $\nu_n \leq N^{2s}$ and $\sum_n \nu_n = \beta_1^{-2} \leq N^{1+s}$. One of the main tasks that will follow is estimating the following sums.

Lemma 5.12 *Given Condition 5.2, $0 < c \leq 10/\sigma$ and $j \leq 3$ there exists an absolute constant C such that*

$$N^{-2s} \delta_2^j e^{-\delta_2 t} \leq \sum_{n=2}^{N} \nu_n \delta_n^j e^{-\delta_n t} \leq C e^{-\delta_2 t} \left(N^{4s} \delta_2^j + N^{1+s} e^{-c\delta_2 t} \right)$$

for N sufficiently large.

Proof For $j \leq 3$,

$$\sum_{n=2}^{N} \nu_n \delta_n^j e^{-\delta_n t} = \left(\sum_{n \in I_c^c} + \sum_{n \in I_c} \right) \nu_n \delta_n^j e^{-\delta_n t}$$

$$\leq \sum_{n \in I_c^c} \nu_n (1+c)^j \delta_2^j e^{-\delta_2 t} + 2^j \sum_{n \in I_c} \nu_n |\lambda_1 - \lambda_n|^j e^{-(1+c)\delta_2 t}.$$

It also follows by Condition 5.2.4 that $\lambda_1 - \lambda_N \leq b_V - a_V + 1$ so that Lemma 5.11 implies for sufficiently large N,

$$\sum_{n=2}^{N} \nu_n \delta_n^j e^{-\delta_n t} \leq C e^{-\delta_2 t} \left(N^{4s} \delta_2^j + N^{1+s} e^{-c\delta_2 t} \right),$$

which is the upper bound we claimed. On the other hand, if we just keep the first term (that should be the largest), we find

$$\sum_{n=2}^{N} \nu_n \delta_n^j e^{-\delta_n t} \geq \nu_2 \delta_2^j e^{-\delta_2 t} \geq N^{-2s} \delta_2^j e^{-\delta_2 t},$$

as was claimed. □

6
Universality for the Toda Algorithm

As noted in the Introduction, in this chapter we consider running the Toda algorithm only until time $T^{(1)}$, the deflation time with block decomposition $k = 1$ fixed, when the norm of the off-diagonal elements in the first row, and hence the first column, is $\mathcal{O}(\epsilon)$. Define

$$E(t) = \sum_{n=2}^{N} |X_{1n}(t)|^2, \qquad (6.1)$$

so that if $E(t) = 0$ then $X_{11}(t)$ is an eigenvalue of H. Thus, with $E(t)$ as in (6.1), the halting time (or 1-deflation time) for the Toda algorithm is given by

$$T^{(1)}(H) = \inf\{t : E(t) \leq \epsilon^2\}. \qquad (6.2)$$

Note that by the min–max principle if $E(t) < \epsilon^2$ then $|X_{11}(t) - \lambda_j| < \epsilon$ for some eigenvalue λ_j of $X(0)$.

The proof of our main result Theorem 6.2, and subsequently Proposition 6.5, proceeds as follows:

- For an initial matrix $H = X(0)$ chosen from some Wigner or Invariant Ensemble, consider the solution $X(t)$ for the Toda flow (3.121) for times $t > 0$.
- Fix the desired accuracy $\epsilon > 0$.
- Express $E(t) = \sum_{n=2}^{N} |X_{1n}(t)|^2$ in terms of the eigenvalues $\{\lambda_j\}$ and first components $\{U_{1j}^+(0)\}$ of the corresponding normalized eigenvectors of the initial matrix $H = X(0)$, as in (3.134), (3.135) and (3.136).
- Solve the scalar equation $E(t) = \epsilon^2$ for the (first) time T such that $E(t) = \epsilon^2$. This is the desired stopping time featured in Theorem 6.2 and Proposition 6.5.

- In this way the proofs of Theorem 6.2 and Proposition 6.5 become problems in "stochastic calculus", i.e., the solution of a single, simple equation $E(t) = \epsilon^2$ for $t = T$, in the case that the coefficients of $E(t)$ are random variables whose precise statistics are derived from the statistics of $H = X(0)$, and presented in Chapter 5.

For invariant as well as generalized Wigner random matrix ensembles, Theorem 5.3 asserts the existence of a constant c_V (which depends on the ensemble) involved in the description of the edge of the equilibrium measure (see (5.2)). Moreover, the following limit exists and only depends on β (either $\beta = 1$ for the real symmetric case or $\beta = 2$ for the complex Hermitian case):

$$F_\beta^{\text{gap}}(t) = \lim_{N \to \infty} \mathbb{P}\left(\frac{1}{c_V^{2/3} 2^{-2/3} N^{2/3} (\lambda_1 - \lambda_2)} \leq t \right), \quad t \geq 0. \qquad (6.3)$$

This limit is discussed further in Definition 5.8. This is the rescaled distribution of the inverse of the top gap in the spectrum of the random matrix. The main result of this monograph is that this distribution is the universal limit of $T^{(1)}$, capturing the fact that the rate of convergence of the Toda algorithm is asymptotically governed by the gap $\lambda_1 - \lambda_2$.

Definition 6.1 (Scaling region) Fix $0 < \sigma < 1$. The **scaling region**[1] for (ϵ, N) is given by $\frac{\alpha}{2} = \frac{\log \epsilon^{-1}}{\log N} \geq 5/3 + \sigma/2$.

Note that for $\epsilon = 10^{-15}$, a relevant value for double-precision arithmetic, (ϵ, N) is in the scaling region for all values of N less than 10^9.

Theorem 6.2 (Universality for $T^{(1)}$) *Let $0 < \sigma < 1$ be fixed and let (ϵ, N) be in the scaling region $\frac{\log \epsilon^{-1}}{\log N} \geq \frac{5}{3} + \frac{\sigma}{2}$. Then if H is distributed according to any real ($\beta = 1$) or complex ($\beta = 2$) invariant or Wigner ensemble, we have*

$$\lim_{N \to \infty} \mathbb{P}\left(\frac{T^{(1)}}{c_V^{2/3} 2^{-2/3} N^{2/3} (\log \epsilon^{-1} - 2/3 \log N)} \leq t \right) = F_\beta^{\text{gap}}(t). \qquad (6.4)$$

Here c_V is the same constant as in (6.3).

Remark 6.3 As noted in Remark 1.4, Theorem 6.2 is valid in the case of invariant ensembles for a very broad class of potentials that are not necessarily convex.

---

[1] From the statement of the theorem, it is reasonable to ask if 5/3 is just an artifact of our method and can be replaced with 2/3 in the definition of the scaling region. Also, one should expect different limits for larger values of ϵ as other eigenvalues will contribute. These questions have yet to be explored.

Example 6.4 Consider the case of real symmetric 2 × 2 matrices. For $X(0) = H$, it follows that as $t \to \infty$, $X_{11}(t) \to \lambda_1$, the largest, or top, eigenvalue, while $X_{22}(t) \to \lambda_2$, the second-largest eigenvalue. And so, one should expect $T^{(1)}$ to be larger for

$$X(0) = H_+ := \begin{bmatrix} -1 & \delta \\ \delta & 1 \end{bmatrix} \quad \text{than for} \quad X(0) = H_- := \begin{bmatrix} 1 & \delta \\ \delta & -1 \end{bmatrix}$$

despite the fact that these matrices have the same eigenvalues. Said differently, it is surprising that the fluctuations of $T^{(1)}$ in Theorem 6.2 depend only on the eigenvalues and are independent of the eigenvectors of H. As in Section 3.7, let $V = (V_{ij})_{1 \le i,j \le 2}$ be the matrix of normalized eigenvectors of $X(0), X(0) = V\Lambda V^*$. It then follows from the calculations in Section 6.2 that

$$|X_{12}(t)|^2 = (\lambda_1 - \lambda_2)^2 (\gamma(t) - \gamma(t)^2),$$

$$\gamma(t) = \frac{|V_{12}(0)|^2 e^{2\lambda_2 t}}{|V_{11}(0)|^2 e^{2\lambda_1 t} + |V_{12}(0)|^2 e^{2\lambda_2 t}}.$$

Thus

$$|X_{12}(t)|^2 \sim (\lambda_1 - \lambda_2)^2 \frac{|V_{12}(0)|^2}{|V_{11}(0)|^2} e^{-2(\lambda_1 - \lambda_2) t}, \text{ as } t \to \infty.$$

First, one should note that this, roughly speaking, explains the appearance of $\lambda_1 - \lambda_2$ in the definition of the universal limit $F_\beta^{\text{gap}}(t)$. Second, a simple calculation shows that as $\delta \downarrow 0$, $|V_{11}(0)| \sim \delta$ for H_+ while $|V_{11}(0)| \sim 1$ for H_-, explaining why $T^{(1)}(H_+) \ge T^{(1)}(H_-)$. However, the matrices H_+ and H_- are not "typical." With high probability, the eigenvectors of random matrices in the ensembles under consideration are delocalized, so that $V_{1j}, j = 1, \ldots, N$ are all of the same order. For general N, we then have $\sum_{k=2}^{N} |X_{1k}|^2 \asymp (\lambda_1 - \lambda_2)^2 e^{-2(\lambda_1 - \lambda_2) t}$ and the dependence on the eigenvectors is effectively removed as $\epsilon \downarrow 0$.

To see that the algorithm computes the top eigenvalue, to an accuracy beyond its fluctuations, we have the following proposition which is a restatement of Theorem 6.22 that shows our error is $O(\epsilon)$ with high probability.

Proposition 6.5 (Computing the largest eigenvalue) *Let (ϵ, N) be in the scaling region. Then if H is distributed according to any real or complex invariant or Wigner ensemble,*

$$\epsilon^{-1} |\lambda_1 - X_{11}(T^{(1)})|$$

converges to zero in probability as $N \to \infty$. Furthermore, both

$$\epsilon^{-1}|b_V - X_{11}(T^{(1)})|, \quad \epsilon^{-1}|\lambda_j - X_{11}(T^{(1)})|$$

converge to ∞ in probability for any $j = j(N) > 1$ as $N \to \infty$, where b_V is the supremum of the support of the equilibrium measure for the ensemble.

The relation of Theorem 6.2 to two-component universality, as discussed in Deift et al. (2014), is the following. Let $\xi = \xi_\beta$ be the random variable with distribution $F_\beta^{\text{gap}}(t)$, $\beta = 1$ or 2. For $\beta = 2$ IEs one can prove that[2]

$$\mathbb{E}[T^{(1)}] = c_V^{2/3} 2^{-2/3} N^{2/3} (\log \epsilon^{-1} - 2/3 \log N) \mathbb{E}[\xi](1 + o(1)), \quad (6.5)$$

$$\sqrt{\text{Var}(T^{(1)})} = \kappa c_V^{2/3} 2^{-2/3} N^{2/3} (\log \epsilon^{-1} - 2/3 \log N)(1 + o(1)) \quad (6.6)$$

for some $\kappa > 0$. By the Law of Large Numbers, if the number of samples is sufficiently large for any fixed, but sufficiently large N, we can restate the result as

$$\mathbb{P}\left(\frac{T^{(1)} - \langle T^{(1)} \rangle}{\sigma_{T^{(1)}}} \leq t\right) \approx F_\beta^{\text{gap}}(\kappa t + \mathbb{E}[\xi]),$$

where $\langle T^{(1)} \rangle$ and $\sigma_{T^{(1)}}$ are the sample average and standard deviation, respectively, as in the introduction. This is a universality theorem for the halting time $T^{(1)}$ as the limiting distribution does not depend on the distribution of the individual entries of the matrix ensemble, just whether it is real or complex.

Remark 6.6 For matrices $H = U \Lambda U^*$, $\Lambda = \text{diag}(\lambda_N, \ldots, \lambda_1)$ where the joint distribution of $\lambda_N \leq \lambda_{N-1} \leq \cdots \leq \lambda_1$ is given by

$$\propto \prod_{j=1}^{N} e^{-N\frac{\beta}{2} V(\lambda_j)} \prod_{j<n} |\lambda_j - \lambda_n|^\beta,$$

and U is distributed (independently) according to Haar measure on either the orthogonal or unitary group, then Theorem 6.2 holds for any $\beta \geq 1$. Here V should satisfy the hypotheses in Definition 5.2.

---

[2] We can also prove (6.5) for $\beta = 1$ IEs. The proofs of these facts require an extension of the level repulsion estimates in Bourgade et al. (2014, Theorem 3.2) to the case "$K = 1$". When $\beta = 2$, again with this extension of Bourgade et al. (2014, Theorem 3.2) to the case "$K = 1$", we can prove that $\kappa = \text{Var}(\xi)$. This extension is known to be true (Bourgade, 2016). The calculations in Table 6.1 are consistent with (6.5) and (6.6) (even for WEs) and lead us to believe that (6.6) also holds for $\beta = 1$. Note that for $\beta = 2$, $\mathbb{E}[\xi^2] < \infty$, but it is believed that $\mathbb{E}[\xi^2] = \infty$ for $\beta = 1$, see Perret and Schehr (2014). In other words, we face the unusual situation where the variance seems to converge, but not to the variance of the limiting distribution.

Remark 6.7 To compute the top eigenvalue of a Hermitian matrix H, one can alternatively consider the flow

$$\partial_t X(t) = HX(t), \quad X(0) = [1, 0, \ldots, 0]^T.$$

It follows that

$$\log \frac{\|X(t+1)\|}{\|X(t)\|} \to \lambda_1, \quad t \to \infty.$$

Define $T_{\text{ODE}}(H) = \inf\left\{t : \left|\log \frac{\|X(t+1)\|}{\|X(t)\|} - \lambda_1\right| \leq \epsilon\right\}$. Using the proof technique we present here, one can show that Theorem 6.2 also holds with $T^{(1)}$ replaced with T_{ODE}. Similar results hold for the direct and inverse power methods, and the QR algorithm without shifts on positive definite random matrices (see Deift and Trogdon, 2017).

6.1 A Numerical Demonstration

We can demonstrate Theorem 6.2 numerically using the following WEs defined by letting X_{ij} for $i < j$ be i.i.d. with distributions:

GUE Mean zero standard complex normal.
BUE $\xi + i\eta$ where ξ and η are each the sum of independent mean zero Bernoulli random variables, i.e., binomial random variables.
GOE Mean zero standard (real) normal.
BOE Mean zero Bernoulli random variable

In each of the above cases, one has to make a choices for the distribution of the diagonal entries. For GOE, GUE we make the standard choice so that they are also invariant ensembles. For BOE and BUE we use standard Bernoulli random variables on the diagonal.

In Figure 6.1, for $\beta = 2$, we show how the histogram of $T^{(1)}$, after rescaling (more precisely, $\tilde{T}^{(1)}$, see (6.7)), matches the density $\partial_t F_2^{\text{gap}}(t)$ which was computed numerically[3] in Witte et al. (2013). In Figure 6.2, for $\beta = 1$, we show the histogram for $T^{(1)}$, after rescaling (again, $\tilde{T}^{(1)}$), matches the density $\partial_t F_1^{\text{gap}}(t)$. To the best of our knowledge, a computationally viable formula for $\partial_t F_1^{\text{gap}}(t)$, analogous to $\partial_t F_2^{\text{gap}}(t)$ in Witte et al. (2013), is not yet known and so we estimate the density $\partial_t F_1^{\text{gap}}(t)$ using Monte Carlo simulations with N large. For convenience, we choose the variance for the

---

[3] Technically, the distribution of the first gap was computed, and then F_2^{gap} can be computed by a change of variables. We thank Folkmar Bornemann for the data to plot F_2^{gap}.

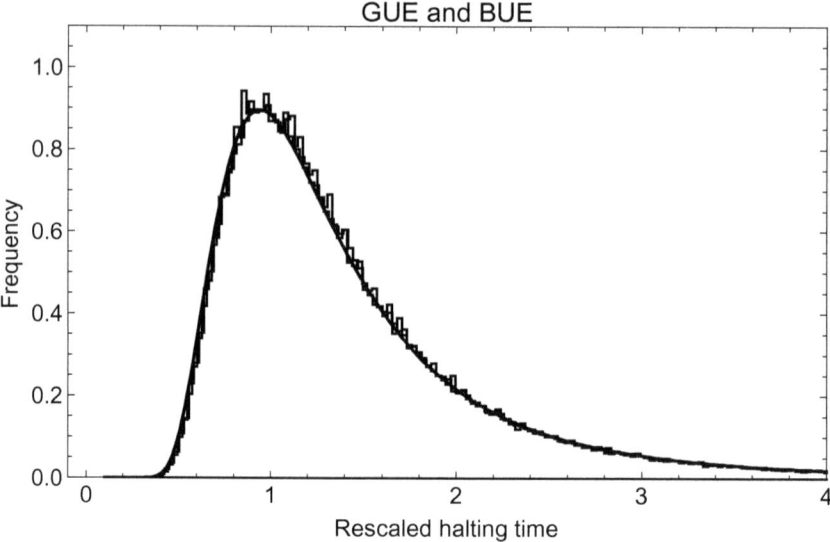

Figure 6.1 The simulated rescaled histogram for $\tilde{T}^{(1)}$ for both BUE and GUE. Here $\epsilon = 10^{-14}$ and $N = 500$ with 250,000 samples. The solid curve is the rescaled density $f_2^{\text{gap}}(t) = \partial_t t F_2^{\text{gap}}(t)$. The density $f_2^{\text{gap}}(t) = \frac{1}{\sigma t^2} A^{\text{soft}}\left(\frac{1}{\sigma t}\right)$, where $A^{\text{soft}}(s)$ is shown in Witte et al. (2013, Figure 1): In order to match the scale in Witte et al. (2013) our choice of distributions (BUE and GUE) we must take $\sigma = 2^{-7/6}$. This is a numerical illustration of Theorem 6.2.

above ensembles so that $[a_V, b_V] = [-2\sqrt{2}, 2\sqrt{2}]$ which, in turn, implies $c_V = 2^{-3/2}$.

It is clear from the proof of Theorem 6.2, where we neglect logarithmically small terms, that the convergence of the left-hand side in (6.4) to F_β^{gap} is slow. In fact, we expect a rate of convergence proportional to $1/\log N$. This means that in order to demonstrate (6.4) numerically with convincing accuracy, one would have to consider very large values of N. In order to display the convergence in (6.4) for more reasonable values of N, we observe, using a simple calculation, that for any fixed $\gamma \neq 0$ the limiting distribution of

$$\tilde{T}^{(1)} = \tilde{T}_\gamma^{(1)} := \frac{T^{(1)}}{c_V^{2/3} 2^{-2/3} N^{2/3} (\log \epsilon^{-1} - 2/3 \log N + \gamma)} \quad (6.7)$$

as $N \to \infty$ is the same as for $\gamma = 0$. A "good" choice for γ is obtained in the following way. To analyze the $T^{(1)}$ in Sections 6.2 and 6.3, we utilize two

6.1 A Numerical Demonstration

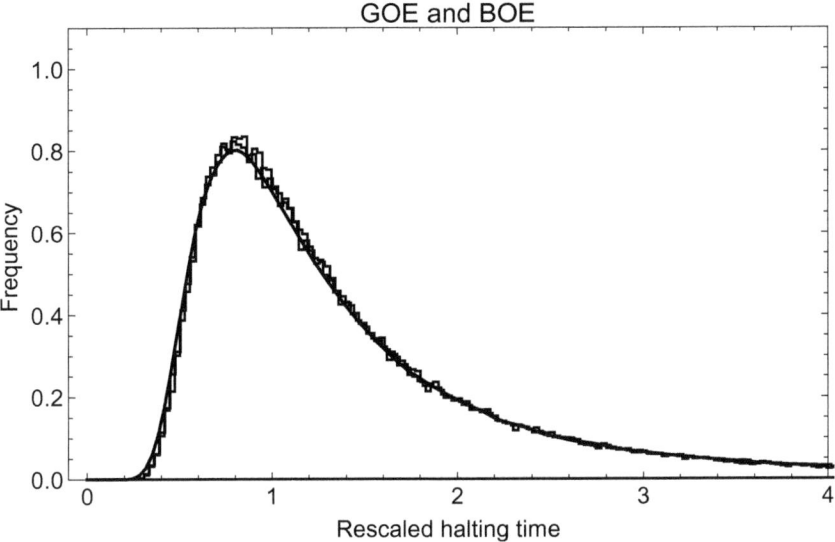

Figure 6.2 The simulated rescaled histogram for $\tilde{T}^{(1)}$ for both BOE and GOE, illustrating Theorem 6.2. Here $\epsilon = 10^{-14}$ and $N = 500$ with 250,000 samples. The solid curve approximates the density $f_1^{\text{gap}}(t) = \partial_t F_1^{\text{gap}}(t)$. We compute $f_1^{\text{gap}}(t)$ by smoothing the histogram for $c_V^{-2/3} 2^{2/3} N^{-2/3} (\lambda_1 - \lambda_2)^{-1}$ when $N = 800$ with 500,000 samples.

approximations to $T^{(1)}$, viz. T^* in (6.17) and $\hat{T}$ in (6.19):

$$T^{(1)} = \hat{T} + (T^{(1)} - T^*) + (T^* - \hat{T}).$$

The parameter γ can be inserted into the calculation by replacing $\hat{T}$ with $\hat{T}_\gamma$

$$\hat{T} \to \hat{T}_\gamma := \frac{(\alpha - 4/3)\log N + 2\gamma}{\delta_2},$$

where γ is chosen to make

$$T^* - \hat{T}_\gamma = \frac{\log N^{2/3}(\lambda_1 - \lambda_2) + \frac{1}{2}\log \nu_2 - \gamma}{\lambda_1 - \lambda_2} \tag{6.8}$$

as small as possible. Here ν_2 and δ_2 are defined at the beginning of Section 5.3. Replacing $\log N^{2/3}(\lambda_1 - \lambda_2)$ and $\log \nu_N$ in (6.8) with the expectation of their respective limiting distributions as $N \to \infty$ (see Theorem 5.7: note that ν_2 is asymptotically distributed as ζ^2 where ζ is Cauchy distributed), we choose

$$\gamma_2 = -\mathbb{E}(\log(c_V^{2/3} 2^{-5/3} \xi_2)) + \frac{1}{2}\mathbb{E}[\log |\zeta|] \approx 0.883 \quad \text{when} \quad \beta = 2,$$

and

$$\gamma_1 = -\mathbb{E}(\log(c_V^{2/3} 2^{-5/3} \xi_1)) + \frac{1}{2}\mathbb{E}[\log|\zeta|] \approx 0.89 \quad \text{when} \quad \beta = 1.$$

Figures 6.1 and 6.2 are plotted using γ_1 and γ_2, respectively.

We can also examine the growth of the mean and standard deviation. We see from Table 6.1 using a million samples and $\epsilon = 10^{-5}$, that the sample standard deviation is on the same order as the sample mean:

$$\sigma_{T^{(1)}} \sim \langle T^{(1)} \rangle \sim N^{2/3}(\log \epsilon^{-1} - 2/3 \log N). \tag{6.9}$$

N	50	100	150	200	250	300
$\log \epsilon^{-1}/\log N - 5/3$	1.28	0.833	0.631	0.506	0.418	0.352
$\langle T^{(1)} \rangle \sigma_{T^{(1)}}^{-1}$ for GUE	1.58	1.62	1.59	1.63	1.6	1.58
$\langle T^{(1)} \rangle \sigma_{T^{(1)}}^{-1}$ for BUE	1.6	1.57	1.6	1.62	1.62	1.58
$\langle T^{(1)} \rangle \sigma_{T^{(1)}}^{-1}$ for GOE	0.506	0.701	0.612	0.475	0.705	0.619
$\langle T^{(1)} \rangle \sigma_{T^{(1)}}^{-1}$ for BOE	0.717	0.649	0.663	0.747	0.63	0.708

Table 6.1 *A numerical demonstration of (6.9). The second row of the table confirms that (ϵ, N) is in the scaling region for, say, $\sigma = 1/2$. The last four rows demonstrate that the ratio of the sample mean to the sample standard deviation is order 1.*

Remark 6.8 The proof of (6.5) for IEs requires the convergence of

$$\mathbb{E}\left[\frac{1}{N^{2/3}(\lambda_1 - \lambda_2)}\right]. \tag{6.10}$$

For BUE, (6.10) must be infinite for all N as there is a nonzero probability that the top two eigenvalues coincide, as the matrix entries are discrete random variables. Nevertheless, the sample mean and sample standard deviation of $T^{(1)}$ are observed to converge, after rescaling. It is an interesting open problem to show that convergence in (6.5) still holds in this case of discrete WEs even though (6.10) is infinite. Specifically, the convergence in the definition of ξ (Definition 5.8) for discrete WEs cannot take place in expectation. Hence $T^{(1)}$ acts as a mollified version of the inverse of the top gap – it is always finite.

6.2 Estimates for the Toda Algorithm

As we have seen in Section 3.6, the Toda flow can be solved explicitly by a QR factorization procedure. Recall that from (3.142), with $m = 1$, one obtains (3.143) and (3.144):

$$E(t) = \sum_{k=2}^{N} |X_{1k}(t)|^2 = \sum_{j=1}^{N} (\lambda_j - X_{11}(t))^2 |V_{1j}(t)|^2,$$

and

$$\lambda_1 - X_{11}(t) = \sum_{j=1}^{N} (\lambda_1 - \lambda_j) |V_{1j}(t)|^2.$$

From these calculations, if $V_{11}(0) \neq 0$, it follows that

$$X_{11}(t) - \lambda_1 \to 0, \quad E(t) \to 0, \quad N \to \infty.$$

While $X_{11}(t) - \lambda_1$ is of course the error in computing λ_1, recall that we use $E(t)$ to determine a convergence criterion as it is easily observable: indeed, as above, if $E(t) < \epsilon$ then $|X_{11}(t) - \lambda_j| < \epsilon^2$ for some j. With high probability, $\lambda_j = \lambda_1$.

Note that, in particular, from the above formulae, $E(t)$ and $\lambda_1 - X_{11}(t)$ depend **only** on the eigenvalues and the moduli of the first components of the eigenvectors of $X(0) = H$. This fact is critical to our analysis. With the notation $\beta_j = |V_{1j}(0)|$, we have that

$$|V_{1j}(t)| = \frac{\beta_j e^{\lambda_j t}}{\left(\sum_{n=1}^{N} \beta_n^2 e^{2\lambda_n t} \right)^{1/2}}.$$

A direct calculation shows that

$$E(t) = E_0(t) + E_1(t),$$

where, in the notation of Section 5.2,

$$E_0(t) = \frac{1}{4} \frac{\sum_{n=2}^{N} \delta_n^2 \nu_n e^{-\delta_n t}}{\left(1 + \sum_{n=2}^{N} \nu_n e^{-\delta_n t} \right)^2}.$$

$$E_1(t) = \frac{\left(\sum_{n=2}^{N} \lambda_n^2 v_n e^{-\delta_n t}\right)\left(\sum_{n=2}^{N} v_n e^{-\delta_n t}\right) - \left(\sum_{n=2}^{N} \lambda_n v_n e^{-\delta_n t}\right)^2}{\left(1 + \sum_{n=2}^{N} v_n e^{-\delta_n t}\right)^2}.$$

Note that $E_1(t) \geq 0$ by the Cauchy–Schwarz inequality; of course, $E_0(t)$ is trivially positive. It follows that $E(t)$ is small if and only if both $E_0(t)$ and $E_1(t)$ are small, a fact that is extremely useful in our analysis.

In terms of the probability measure $\rho_{N,k}$ on subsets S of $\{k, k+1, \ldots, N\}$ defined by

$$\rho_{N,k}(S) = \left(\sum_{n=k}^{N} v_n e^{-\delta_n t}\right)^{-1} \sum_{n \in S} v_n e^{-\delta_n t},$$

and the function $\lambda(j) = \lambda_j$

$$E(t) = \mathrm{Var}_{\rho_{N,1}}(\lambda).$$

We will also use the alternate expression

$$E_1(t) = \left(\frac{\sum_{n=2}^{N} v_n e^{-\delta_n t}}{1 + \sum_{n=2}^{N} v_n e^{-\delta_n t}}\right)^2 \mathrm{Var}_{\rho_{N,2}}(\lambda). \quad (6.11)$$

Additionally,

$$\lambda_1 - X_{11}(t) = \frac{1}{2} \frac{\sum_{n=2}^{N} \delta_n \beta_n^2 e^{\delta_n t}}{1 + \sum_{n=2}^{N} \beta_n^2 e^{\delta_n t}}. \quad (6.12)$$

6.2.1 The Halting Time and Its Approximation

To aid the reader we provide a glossary to summarize inequalities for parameters and quantities that have previously appeared:

(1) $0 < \sigma < 1$ is fixed,
(2) $0 < p < 1/3$,
(3) $\alpha \geq 10/3 + \sigma$,

(4) $s \leq \min\{\sigma/44, p/8\}$,
(5) $\alpha - 4/3 - 44s \geq 2$,
(6) c can be chosen for convenience line by line when estimating sums with Lemma 5.12, provided that $c \leq 10/\sigma$,
(7) $\delta_n = 2(\lambda_1 - \lambda_n)$,
(8) $\nu_n = \beta_n^2/\beta_1^2$,
(9) given Condition 5.2,
- $2N^{-2/3-s} \leq \delta_2 \leq 2N^{-2/3+s}$,
- $N^{-2s} \leq \nu_n \leq N^{2s}$,
- $\sum_{n=j}^{N} \nu_n \leq \sum_{n=1}^{N} \nu_n = \beta_1^{-2} \leq N^{1+s}$, for $1 \leq j \leq N$, and

(10) $C > 0$ is a generic constant.

Definition 6.9 Recall the halting time (or the 1-deflation time) for the Toda lattice is defined to be, see (6.2),

$$T^{(1)} = \inf\{t : E(t) \leq \epsilon^2\}.$$

We find bounds on the halting time.

Lemma 6.10 *Under Condition 5.2, the halting time $T^{(1)}$ for the Toda lattice satisfies*

$$(\alpha - 4/3 - 5s)\log N/\delta_2 \leq T^{(1)} \leq (\alpha - 4/3 + 7s)\log N/\delta_2$$

for sufficiently large N.

Proof We use that $E(t) \geq E_0(t)$ so if $E_0(t) > N^{-\alpha}$ then $T^{(1)} \geq t$. First, we show that $E_0(t) > \epsilon^2$, $0 \leq t \leq \sigma/2 \log N/\delta_2$ and sufficiently large N and then we use this to show that $E_0(t) > \epsilon^2$, $t \leq (\alpha - 4/3 - 5s)\log N/\delta_2$ and sufficiently large N.

Indeed, consider $t = a \log N/\delta_2$ for $0 \leq a \leq \sigma/2$. Using Lemma 5.12,

$$1 + \sum_{n=2}^{N} \nu_n e^{-\delta_n t} \leq 1 + Ce^{-\delta_2 t}\left(N^{4s} + N^{1+s}e^{-c\delta_2 t}\right). \quad (6.13)$$

Then using Lemma 5.12 again we have

$$E_0(t) \geq N^{-2s}\delta_2^2 e^{-\delta_2 t}\left(1 + Ce^{-\delta_2 t}\left(N^{4s} + N^{1+s}e^{-c\delta_2 t}\right)\right)^{-2}.$$

Since $a \leq \sigma/2$, we find

$$E_0(t) \geq N^{-4s-4/3-\sigma/2}\left(1 + C(N^{4s} + N^{1+s})\right)^{-2} \geq CN^{-8s-10/3-\sigma/2}$$

for some new constant $C > 0$. This last inequality follows because $N^{4s} \leq$

N^{1+s} as $s \leq 1/44$ (see Condition 5.2). But then from Definition 6.1 this right-hand side is larger than $\epsilon^2 = N^{-\alpha}$ for sufficiently large N. Now, consider $t = a \log N/\delta_2$ for $\sigma/2 \leq a \leq (\alpha - 4/3 - 5s)$. We choose $c = 2(2+s)/\sigma \leq 10/\sigma$

$$E_0(t) \geq \frac{1}{4} N^{-4s-4/3-a} \left(1 + C(N^{4s-a} + N^{1+s-ca})\right)^{-2}$$
$$\geq N^{-\alpha+s}(1 + C(N^{4s-\sigma/2} + N^{-1})) > N^{-\alpha}$$

for sufficiently large N. Here we used that $s \leq \sigma/44$. This shows $(\alpha - 4/3 - 5s) \log N/\delta_2 \leq T^{(1)}$ for N sufficiently large.

Now, we work on the upper bound. Let $t = a \log N/\delta_2$ for $a \geq (\alpha - 4/3 + 7s)$ and we find, using Lemma 5.12,

$$E_0(t) \leq CN^{-a}\left(N^{-4/3+6s} + N^{1+s-ca}\right).$$

Then using the minimum value for a,

$$E_0(t) \leq N^{-\alpha}\left(C(N^{-s} + CN^{1+7s-ca+4/3})\right).$$

By property (5) above $a > 2$. If we set $c = 3$ and use $s \leq 1/44$, then $1 + 7s - ca + 4/3 \leq -5 + 4/3 + 7s \leq -2$

$$E_0(t) \leq N^{-\alpha}\left(C(N^{-s} + CN^{-2})\right) < CN^{-\alpha-s}$$

for sufficiently large N.

Next, we must estimate $E_1(t)$ when $a \geq (\alpha - 4/3 + 7s)$. We use (6.11) and $\text{Var}_{\rho_{N,2}}(\lambda) \leq C$. Then by (6.13),

$$E_1(t) \leq CN^{-2a}(N^{4s} + N^{1+s-ca})^2.$$

Again, using $c = 1$ and the fact that $a > 2$, we have

$$E_1(t) \leq CN^{-\alpha}N^{8s-\alpha+8/3-14s} \leq CN^{-\alpha}N^{-\alpha+8/3} \leq N^{-\alpha} \tag{6.14}$$

for N sufficiently large. This shows $T^{(1)} \leq (\alpha - 4/3 + 7s) \log N/\delta_2$ for sufficiently large N as $E(t) = E_0(t) + E_1(t) \leq \epsilon^2$ for some time

$$t < (\alpha - 4/3 + 7s) \log N/\delta_2$$

and N is sufficiently large. □

In light of this lemma we define the interval

$$L_\alpha = [(\alpha - 4/3 - 5s) \log N/\delta_2, (\alpha - 4/3 + 7s) \log N/\delta_2].$$

6.2 Estimates for the Toda Algorithm

Next, we estimate the derivative of $E_0(t)$. We find

$$E_0'(t) = \frac{-\left(\sum_{n=2}^{N} \delta_n^3 \nu_n e^{-\delta_n t}\right)\left(1 + \sum_{n=2}^{N} \nu_n e^{-\delta_n t}\right) + 2\left(\sum_{n=2}^{N} \delta_n^2 \nu_n e^{-\delta_n t}\right)\left(\sum_{n=2}^{N} \delta_n \nu_n e^{-\delta_n t}\right)}{\left(1 + \sum_{n=2}^{N} \nu_n e^{-\delta_n t}\right)^3}. \tag{6.15}$$

Lemma 6.11 *Under Condition 5.2 and $t \in L_\alpha$,*

$$-E_0'(t) \geq CN^{-12s-\alpha-2/3}$$

for sufficiently large N.

Proof We use (6.15). The denominator is bounded below by unity so we estimate the numerator. By Lemma 5.12,

$$\left(\sum_{n=2}^{N} \delta_n^3 \nu_n e^{-\delta_n t}\right)\left(1 + \sum_{n=2}^{N} \nu_n e^{-\delta_n t}\right) \geq \sum_{n=2}^{N} \delta_n^3 \nu_n e^{-\delta_n t} \geq N^{-2s} \delta_2^3 e^{-\delta_2 t}.$$

For $t \in L_\alpha$,

$$N^{-2s} \delta_2^3 e^{-\delta_2 t} \geq N^{-12s-2/3-\alpha}.$$

Next, again by Lemma 5.12,

$$\left(\sum_{n=2}^{N} \delta_n \nu_n e^{-\delta_n t}\right)\left(\sum_{n=2}^{N} \delta_n^2 \nu_n e^{-\delta_n t}\right)$$

$$\leq Ce^{-2\delta_2 t}\left(N^{4s}\delta_2^2 + N^s e^{-c\delta_2 t}\right)\left(N^{4s}\delta_2 + N^{1+s} e^{-c\delta_2 t}\right).$$

The estimate with $c = 2$ is

$$N^{4s}\delta_2^2 + N^s e^{-c\delta_2 t} \leq 4N^{6s-4/3} + N^{s-4} \leq CN^{6s-4/3},$$
$$N^{4s}\delta_2 + N^s e^{-c\delta_2 t} \leq 2N^{5s-2/3} + N^{s-4} \leq CN^{5s-2/3},$$

where we used $t \geq 2\log N/\delta_2$ and $s \leq 1/44$. Further,

$$e^{-2\delta_2 t} \leq N^{-\alpha} N^{8/3-\alpha+10s} \leq N^{-\alpha-2/3-\sigma+10s}$$

as $s \leq \sigma/44$. Then

$$-E_0'(t) \geq N^{-12s-2/3-\alpha} - CN^{-\alpha-2/3-\sigma+10s},$$

and hence
$$-E_0'(t) \geq N^{-12s-2/3-\alpha}(1 - CN^{-\sigma+22s}) \geq 0$$
for N sufficiently large as $s \leq \sigma/44$. □

Now we look at the leading-order behavior of $E_0(t)$:
$$E_0(t) = \frac{1}{4}\delta_2^2 v_2 e^{-\delta_2 t} \cdot \frac{1 + \sum_{n=3}^{N} \frac{\delta_n^2}{\delta_2^2} \frac{v_n}{v_2} e^{-(\delta_n - \delta_2)t}}{\left(1 + \sum_{n=2}^{N} v_n e^{-\delta_n t}\right)^2}. \tag{6.16}$$

Define T^* by
$$\frac{1}{4}\delta_2^2 v_2 e^{-\delta_2 T^*} = N^{-\alpha},$$
$$T^* = \frac{\alpha \log N + 2\log \delta_2 + \log v_2 - 2\log 2}{\delta_2}. \tag{6.17}$$

Lemma 6.12 *Under Condition 5.2,*
$$(\alpha - 4/3 - 4s)\log N/\delta_2 \leq T^* \leq (\alpha - 4/3 + 4s)\log N/\delta_2$$
for sufficiently large N.

Proof This follows immediately from the statements
$$N^{-2s} \leq v_2 \leq N^{2s},$$
$$2N^{-2/3-s} \leq \delta_2 \leq 2N^{-2/3+s}. \qquad \square$$

Thus, given Condition 5.2, $T^* \in L_\alpha$. The quantity that we want to estimate is $N^{-2/3}|T^{(1)} - T^*|$. And we do this by considering the formula
$$E_0(T^{(1)}) - E_0(T^*) = E_0'(\eta)(T^{(1)} - T^*) \quad \text{for some } \eta \in L_\alpha.$$

And because E_0 is monotone in L_α, $E_0(T^{(1)}) = E(T^{(1)}) - E_1(T^{(1)}) = N^{-\alpha} - E_1(T^{(1)})$, we have
$$|T^{(1)} - T^*| \leq \frac{|N^{-\alpha} - E_0(T^*) - E_1(T^{(1)})|}{\min_{\eta \in L_\alpha} |E_0'(\eta)|}$$
$$\leq \frac{|N^{-\alpha} - E_0(T^*)| + \max_{\eta \in L_\alpha} |E_1(\eta)|}{\min_{\eta \in L_\alpha} |E_0'(\eta)|}. \tag{6.18}$$

See Figure 6.3 for a schematic of $E_0, E, T^{(1)}$ and T^*.

6.2 Estimates for the Toda Algorithm

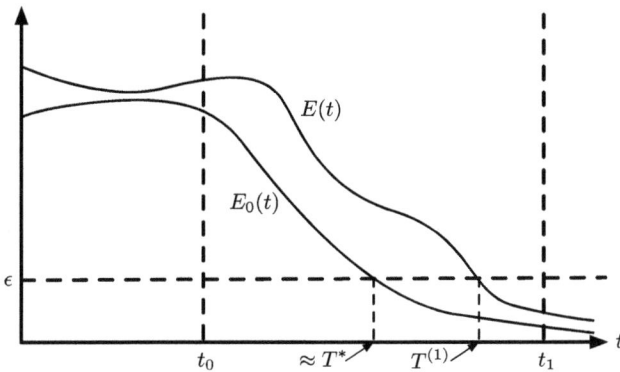

Figure 6.3 A schematic for the relationship between the functions $E_0(t)$, $E(t)$ and the times $T^{(1)}$ and T^*. Here $t_0 = (\alpha - 4/3 - 5s)\log N/\delta_2$ and $t_1 = (\alpha - 4/3 + 7s)\log N/\delta_2$. Note that E_0 is monotone on $[t_0, t_1]$.

Since we already have an adequate estimate on $E_1(T)$ in (6.14), it remains to estimate $|N^{-\alpha} - E_0(T^*)|$.

Lemma 6.13 *Given Conditions 5.1 and 5.2,*

$$|E_0(T^*) - N^{-\alpha}| \le CN^{-\alpha - 2p + 4s}.$$

Proof From (6.16) and (6.17), we obtain

$$|E_0(T^*) - N^{-\alpha}|$$

$$= N^{-\alpha} \frac{\left| \sum_{n=3}^{N} \frac{\delta_n^2}{\delta_2^2} \frac{v_n}{v_2} e^{-(\delta_n - \delta_2)T^*} - 2 \sum_{n=2}^{N} v_n e^{-\delta_n T^*} - \left(\sum_{n=2}^{N} v_n e^{-\delta_n T^*} \right)^2 \right|}{\left(1 + \sum_{n=2}^{N} v_n e^{-\delta_n T^*} \right)^2}.$$

We estimate the terms in the numerator individually using the bounds on T^*. For $c = 1$, we use that $\alpha - 4/3 - 4s > 2$ and Lemma 5.12 to find

$$\sum_{n=2}^{N} v_n e^{-\delta_n T^*} \le CN^{-\alpha + 4/3 + 4s}\left(N^{4s} + N^{1+s-2c}\right) \le CN^{-2-\sigma+8s} \le N^{-2}$$

for sufficiently large N. Then we consider the first term in the numerator using the index set I_c and Condition 5.1. We define $\hat{I}_c = I_c \cap \{3, \ldots, N\}$ and $\hat{I}_c^c$ to denote the complement relative to $\{3, \ldots, N\}$. Continuing,

$$\ell(T^*) := \sum_{n=3}^{N} \frac{\delta_n^2}{\delta_2^2} \frac{\nu_n}{\nu_2} e^{-(\delta_n-\delta_2)T^*} = \left(\sum_{n \in \hat{I}_c^c} + \sum_{n \in \hat{I}_c} \right) \frac{\delta_n^2}{\delta_2^2} \frac{\nu_n}{\nu_2} e^{-(\delta_n-\delta_2)T^*}.$$

For $n \in \hat{I}_c^c$, $\delta_n^2/\delta_2^2 \leq (1+c)^2$,

$$\delta_n - \delta_2 = 2(\lambda_2 - \lambda_n) \geq 2(\lambda_2 - \lambda_3) \geq p\delta_2,$$

using Condition 5.1. On the other hand, for $n \in \hat{I}_c$, $\delta_n > (1+c)\delta_2$, and if $c = 3$,

$$\delta_n - \delta_2 > c\delta_2 = p\delta_2 + (c-p)\delta_2 \geq p\delta_2 + 2\delta_2,$$

as $p < 1/3$ and hence $c > 2 + p$. Using Lemma 5.11 to estimate $|\hat{I}_c^c|$, we obtain

$$\sum_{n \in \hat{I}_c^c} \frac{\delta_n^2}{\delta_2^2} \frac{\nu_n}{\nu_2} e^{-(\delta_n-\delta_2)T^*} \leq (1+c)^2 N^{4s} e^{-p\delta_2 T^*},$$

$$\sum_{n \in \hat{I}_c} \frac{\delta_n^2}{\delta_2^2} \frac{\nu_n}{\nu_2} e^{-(\delta_n-\delta_2)T^*} \leq [\max_n \delta_n^2] N^{7/3+3s} e^{-(p+2)\delta_2 T^*}.$$

Given Condition 5.2 $[\max_n \delta_n^2] \leq 4(b_V - a_V + 1)^2$ and hence for some $C > 0$, using that $\alpha - 4/3 - 4s > 2$,

$$\ell(T^*) \leq C e^{-p\delta_2 T^*} \left(N^{4s} + N^{7/3+3s} e^{-2\delta_2 T^*} \right)$$
$$\leq C N^{-p(\alpha-4/3-4s)} (N^{4s} + N^{7/3+3s-2(\alpha-4/3-4s)})$$
$$\leq C N^{-p(\alpha-4/3-4s)} (N^{4s} + N^{-5/3+3s})$$
$$\leq C N^{-2p+4s} (1 + N^{-5/3-s}).$$

Thus,

$$\ell(T^*) \leq C N^{-2p+4s}.$$

From this it follows that

$$|E_0(T^*) - N^{-\alpha}| \leq C N^{-\alpha-2p+4s}. \quad \square$$

Lemma 6.14 *For σ, p fixed, $s < \min\{\sigma/44, p/8\}$ and Conditions 5.1 and 5.2,*

$$N^{-2/3} |T^{(1)} - T^*| \leq C \left(N^{-2p+16s} + N^{-\sigma-2/3+12s} \right) \to 0, \quad \text{as } N \to \infty.$$

Proof Combining Lemmas 6.11 and 6.13 with (6.14), we have

$$E_1(t) \leq C N^{-\alpha-2/3-\sigma}.$$

6.3 Adding Probability

Together with (6.18) we obtain, sufficiently large N,

$$N^{-2/3}|T^{(1)} - T^*| \leq CN^{-2/3}N^{\alpha+12s+2/3}\left(N^{-\alpha-2p+4s} + N^{-\alpha}N^{-2/3-\sigma}\right)$$
$$\leq C\left(N^{-2p+16s} + N^{-\sigma-2/3+12s}\right). \qquad \square$$

From (6.12), we have

$$|\lambda_1 - X_{11}(t)| = \frac{1}{2}\frac{\sum_{n=2}^{N}\delta_n v_n e^{-\delta_n t}}{1+\sum_{n=2}^{N}v_n e^{-\delta_n t}} \leq \frac{1}{2}\sum_{n=2}^{N}\delta_n v_n e^{-\delta_n t}.$$

Lemma 6.15 *Given Condition 5.2, σ and p fixed and $s < \min\{\sigma/44, p/8\}$,*

$$\varepsilon^{-1}|\lambda_1 - X_{11}(T^{(1)})| = N^{\alpha/2}|\lambda_1 - X_{11}(T^{(1)})| \leq CN^{-1}$$

for sufficiently large N.

Proof We use Lemma 5.12 with $c = 1$. By 6.10 we have

$$|\lambda_1 - X_{11}(T^{(1)})| \leq CN^{-\alpha+4/3+5s}(N^{-2/3+5s} + N^{-1+s}) \leq CN^{-\alpha/2}N^{-1}$$

because $\alpha \geq 10/3 + \sigma$. $\qquad \square$

6.3 Adding Probability

We now use the probabilistic facts about Conditions 5.2 and 5.1 as stated in Theorems 5.9 and 5.10 to understand $T^{(1)}$ and T^* as random variables.

Lemma 6.16 *For $\alpha \geq 10/3 + \sigma$ and $\sigma > 0$,*

$$\frac{|T^{(1)} - T^*|}{N^{2/3}}$$

converges to zero in probability as $N \to \infty$.

Proof Let $\varepsilon > 0$. Then

$$\mathbb{P}\left(\frac{|T^{(1)} - T^*|}{N^{2/3}} > \varepsilon\right) = \mathbb{P}\left(\frac{|T^{(1)} - T^*|}{N^{2/3}} > \varepsilon, G_{N,p} \cap R_{N,s}\right)$$
$$+ \mathbb{P}\left(\frac{|T^{(1)} - T^*|}{N^{2/3}} > \varepsilon, G_{N,p}^c \cup R_{N,s}^c\right).$$

If s satisfies the hypotheses in Lemma 6.14, $s < \min\{\sigma/44, p/8\}$, then on the set $G_{N,p} \cap R_{N,s}$, $N^{-2/3}|T^{(1)} - T^*| < \varepsilon$ for N sufficiently large, and hence

$$\mathbb{P}\left(\frac{|T^{(1)} - T^*|}{N^{2/3}} > \varepsilon, G_{N,p} \cap R_{N,s}\right) \to 0,$$

as $N \to \infty$. We then estimate
$$\mathbb{P}\left(\frac{|T^{(1)} - T^*|}{N^{2/3}} > \varepsilon, G_{N,p}^c \cup R_{N,s}^c\right) \leq \mathbb{P}(G_{N,p}^c) + \mathbb{P}(R_{N,s}^c),$$
and by Theorem 5.9,
$$\limsup_{N \to \infty} \mathbb{P}\left(\frac{|T^{(1)} - T^*|}{N^{2/3}} > \varepsilon, G_{N,p}^c \cup R_{N,s}^c\right) \leq \limsup_{N \to \infty} \mathbb{P}(G_{N,p}^c).$$
This is true for any $0 < p < 1/3$, and we use Theorem 5.10. So, as $p \downarrow 0$, we find
$$\lim_{N \to \infty} \mathbb{P}\left(\frac{|T^{(1)} - T^*|}{N^{2/3}} > \varepsilon\right) = 0.$$

$\square$

Define
$$\hat{T} = \frac{(\alpha - 4/3) \log N}{\delta_2}. \tag{6.19}$$
We need the following simple lemmas in what follows.

Lemma 6.17 *If $X_N \to X$ in distribution[4] as $N \to \infty$ then*
$$\mathbb{P}(|X_N/a_N| < 1) = 1 + o(1),$$
as $N \to \infty$ provided that $a_N \to \infty$.

Proof For two points of continuity a, b of $F(t) = \mathbb{P}(X \leq t)$, we have
$$\mathbb{P}(a < X_N \leq b) \to \mathbb{P}(a < X \leq b).$$
Let $M > 0$ such that $\pm M$ is a point of continuity of F. Then for sufficiently large N, $a_N > M$ and
$$\liminf_{N \to \infty} \mathbb{P}(-a_N < X_N < a_N) \geq \liminf_{N \to \infty} \mathbb{P}(-M < X_N \leq M)$$
$$= \mathbb{P}(-M < X \leq M).$$
Letting $M \to \infty$, we see that $\mathbb{P}(-a_N \leq X_N \leq a_N) = 1 + o(1)$ as $N \to \infty$. $\square$

Letting $a_N \to \eta a_N$, $\eta > 0$, we see that the following is true.

Corollary 6.18 *If $X_N \to X$ in distribution as $N \to \infty$ then*
$$|X_N/a_N|$$
converges to zero in probability provided $a_N \to \infty$.

---
[4] For convergence in distribution, we require that the limiting random variable X satisfies $\mathbb{P}(|X| < \infty) = 1$.

6.3 Adding Probability

Lemma 6.19 *If as $N \to \infty$, $X_N \to X$ in distribution and $|X_N - Y_N| \to 0$ in probability then $Y_N \to X$ in distribution.*

Proof Let t be a point of continuity for $\mathbb{P}(X \leq t)$, then for $\varepsilon > 0$,

$$\mathbb{P}(Y_N \leq t) = \mathbb{P}(Y_N \leq t, X_N \leq t + \varepsilon) + \mathbb{P}(Y_N \leq t, X_N > t + \varepsilon)$$
$$\leq \mathbb{P}(X_N \leq t + \varepsilon) + \mathbb{P}(Y_N - X_N \leq t - X_N, t - X_N < -\varepsilon)$$
$$\leq \mathbb{P}(X_N \leq t + \varepsilon) + \mathbb{P}(|Y_N - X_N| > \varepsilon).$$

Interchanging the roles of X_N and Y_N and replacing t with $t - \varepsilon$, we find

$$\mathbb{P}(X_N \leq t - \varepsilon) \leq \mathbb{P}(Y_N \leq t) + \mathbb{P}(|Y_N - X_N| > \varepsilon)$$

$$\leq \mathbb{P}(X_N \leq t + \varepsilon) + 2\mathbb{P}(|Y_N - X_N| > \varepsilon).$$

From this we find that for any ε such that $t \pm \varepsilon$ are points of continuity,

$$\mathbb{P}(X \leq t - \varepsilon) \leq \liminf_{N \to \infty} \mathbb{P}(Y_N \leq t) \leq \limsup_{N \to \infty} \mathbb{P}(Y_N \leq t) \leq \mathbb{P}(X \leq t + \varepsilon).$$

By sending $\varepsilon \downarrow 0$ the result follows. $\square$

Now, we compare T^* to $\hat{T}$.

Lemma 6.20 *For $\alpha \geq 10/3 + \sigma$,*

$$\frac{|T^* - \hat{T}|}{N^{2/3} \log N}$$

converges to zero in probability as $N \to \infty$.

Proof Consider

$$\frac{T^* - \hat{T}}{N^{2/3} \log N} = \frac{1}{\log N} \frac{\log \nu_2 + 2\log N^{2/3} \delta_2}{N^{2/3} \delta_2}$$
$$= \frac{1}{\sqrt{\log N}} \left(\frac{1}{(\log N)^{1/4}} |N^{2/3} \delta_2|^{-1} \right) \left(\frac{2}{(\log N)^{1/4}} \log \nu_2 + \frac{1}{(\log N)^{1/4}} \log N^{2/3} \delta_2 \right).$$

For

$$L_N = \left\{ \frac{1}{(\log N)^{1/4}} |N^{2/3} \delta_2|^{-1} \leq 1 \right\},$$

$$U_N = \left\{ \frac{1}{(\log N)^{1/4}} |\log \nu_2| \leq 1 \right\},$$

$$P_N = \left\{ \frac{1}{(\log N)^{1/4}} |\log N^{2/3} \delta_2| \leq 1 \right\},$$

we have $\mathbb{P}(L_N^c) + \mathbb{P}(U_N^c) + \mathbb{P}(P_N^c) \to 0$ as $N \to \infty$ by Lemma 6.17 and Theorem 5.7. For these calculations, it is important that the limiting distribution function for $N^{2/3}\delta_2$ is continuous at zero, see Theorem 5.7. Then for $\varepsilon > 0$,

$$\mathbb{P}\left(\left|\frac{T^* - \hat{T}}{N^{2/3} \log N}\right| > \varepsilon\right) = \mathbb{P}\left(\left|\frac{T^* - \hat{T}}{N^{2/3} \log N}\right| > \varepsilon, L_N \cap U_N \cap P_N\right) \\ + \mathbb{P}\left(\left|\frac{T^* - \hat{T}}{N^{2/3} \log N}\right| > \varepsilon, L_N^c \cup U_N^c \cup P_N^c\right). \quad (6.20)$$

On the set $L_N \cap U_N \cap P_N$, we estimate

$$\left|\frac{T^* - \hat{T}}{N^{2/3} \log N}\right| \le \frac{3}{\sqrt{\log N}}.$$

Hence, the first term on the right-hand side of (6.20) is zero for sufficiently large N and the second term is bounded by $\mathbb{P}(U_N^c) + \mathbb{P}(L_N^c) + \mathbb{P}(P_N^c)$ which tends to zero. This shows convergence in probability. □

We now arrive at our main result.

Theorem 6.21 *If $\alpha \ge 10/3 + \sigma$ and $\sigma > 0$ then*

$$\lim_{N\to\infty} \mathbb{P}\left(\frac{2^{2/3} T^{(1)}}{c_V^{2/3}(\alpha - 4/3)N^{2/3} \log N} \le t\right) = F_\beta^{\text{gap}}(t).$$

Proof Combining Lemma 6.16 and Lemma 6.20, we have that

$$\left|2^{2/3} \frac{T^{(1)} - \hat{T}}{c_V^{2/3}(\alpha - 4/3)N^{2/3} \log N}\right|$$

converges to zero in probability. Then by Lemma 6.19 and Theorem 5.7, the result follows as

$$\lim_{N\to\infty} \mathbb{P}\left(\frac{2^{2/3} \hat{T}}{c_V^{2/3}(\alpha - 4/3)N^{2/3} \log N} \le t\right) \\ = \lim_{N\to\infty} \mathbb{P}(c_V^{-2/3} 2^{2/3} N^{-2/3}(\lambda_1 - \lambda_2)^{-1} \le t) = F_\beta^{\text{gap}}(t). \quad \square$$

We also prove a result concerning the true error $|\lambda_1 - X_{11}(T^{(1)})|$:

Proposition 6.22 *For $\alpha \ge 10/3 + \sigma$ and $\sigma > 0$ and any $q < 1$,*

$$N^{\alpha/2+q}|\lambda_1 - X_{11}(T^{(1)})|$$

converges to zero in probability as $N \to \infty$. Furthermore, for any $r > 0$,

$$N^{2/3+r}|\gamma_1 - X_{11}(T^{(1)})|, \quad N^{2/3+r}|\lambda_j - X_{11}(T^{(1)})|,$$

converges to ∞ in probability, if $j = j(N) > 1$.

Proof We recall that $R_{N,s}$ is the set on which Condition 5.2 holds. Then for any $\eta > 0$

$$\mathbb{P}(N^{\alpha/2+q}|\lambda_1 - X_{11}(T^{(1)})| > \eta)$$
$$= \mathbb{P}(N^{\alpha/2+q}|\lambda_1 - X_{11}(T^{(1)})| > \eta, R_{N,s})$$
$$+ \mathbb{P}(N^{\alpha/2+q}|\lambda_1 - X_{11}(T^{(1)})| > \eta, R_{N,s}^c)$$
$$\leq \mathbb{P}(N^{\alpha/2+q}|\lambda_1 - X_{11}(T^{(1)})| > \eta, R_{N,s}) + \mathbb{P}(R_{N,s}^c).$$

Using Lemma 6.15, the first term on the right-hand side is zero for sufficiently large N and the second term vanishes from Theorem 5.9. This shows the first statement, i.e.,

$$\lim_{N \to \infty} \mathbb{P}(N^{\alpha/2+q}|\lambda_1 - X_{11}(T^{(1)})| > \eta) = 0.$$

For the second statement, on the set $R_{N,s}$ with $s < \min\{r, \sigma/44, p/8\}$,

$$|\lambda_j - X_{11}(T^{(1)})| \geq |\lambda_j - \lambda_1| - |\lambda_1 - X_{11}(T^{(1)})| \geq |\lambda_2 - \lambda_1| - |\lambda_1 - X_{11}(T^{(1)})|,$$

and for sufficiently large N (see Lemma 6.15),

$$N^{2/3+r}|\lambda_j - X_{11}(T^{(1)})| \geq N^r(N^{2/3}|\lambda_2 - \lambda_1| - N^{-1/3-\alpha/2})$$
$$\geq N^{r-s}(1 - CN^{-1/3-\alpha/2+s}).$$

This tends to ∞ as $s < 1/3$ and $s < r$. Hence for any $K > 0$, again using the arguments of Theorem 6.21,

$$\mathbb{P}\left(N^{2/3+r}|\lambda_j - X_{11}(T^{(1)})| > K\right) = \mathbb{P}\left(N^{2/3+r}|\lambda_j - X_{11}(T^{(1)})| > K, R_{N,s}\right)$$
$$+ \mathbb{P}\left(N^{2/3+r}|\lambda_j - X_{11}(T^{(1)})| > K, R_{N,s}^c\right).$$

For sufficiently large N, the first term on the right-hand side is equal to $\mathbb{P}(R_{N,s})$ and the second term is bounded by $\mathbb{P}(R_{N,s}^c)$ and hence

$$\lim_{N \to \infty} \mathbb{P}\left(N^{2/3+r}|\lambda_j - X_{11}(T^{(1)})| > K\right) = 1.$$

Next, under the same assumption (Condition 5.2),

$$N^{2/3+r}|b_V - X_{11}(T^{(1)})| \geq N^r(N^{2/3}|b_V - \lambda_1| - CN^{-1/3-\alpha/2}).$$

From Corollary 6.18 and Theorem 5.7 using $\gamma_N = b_V$,

$$N^{-r}(N^{2/3}|b_V - \lambda_1| - CN^{-1/3-\alpha/2})^{-1}$$

converges to zero in probability (with no point mass at zero), implying its inverse converges to ∞ in probability. This shows $N^\alpha|b_V - X_{11}(T^{(1)})|$ converges to ∞ in probability. □

6.3.1 $T^{(1)}$ when $\epsilon = O(1)$

We close this monograph with a short discussion of $T^{(1)}$ in different regimes. In light of Theorem 3.35, we define the probability measure

$$\mu_{N,t}(d\lambda) = \sum_{j=1}^{N} |V_{1j}(t)|^2 \delta_{\lambda_j}(d\lambda).$$

Then

$$X_{11}(t) = \mathbb{E}_{\mu_{N,t}}[\lambda],$$

and

$$E(t) = \text{Var}_{\mu_{N,t}}[\lambda].$$

For many choices of random matrix initial data, IEs and WEs included, one has

$$\mu_{N,0}(d\lambda) \to \rho(\lambda)d\lambda.$$

Thus, local laws for random matrices (Bloemendal et al., 2014; Knowles and Yin, 2017) imply such a limit and give the sense in which the limit exists. Then, for fixed t, one expects

$$\mu_{N,t}(d\lambda) \to \frac{e^{\lambda t}\rho(\lambda)d\lambda}{\int e^{\lambda t}\rho(\lambda)d\lambda}.$$

And therefore, as $N \to \infty$, for t fixed,

$$X_{11}(t) \to m(t) := \frac{\int \lambda e^{\lambda t}\rho(\lambda)d\lambda}{\int e^{\lambda t}\rho(\lambda)d\lambda}, \qquad (6.21)$$

$$E(t) \to E_{\infty}(t) := \frac{\int (\lambda - m(t))^2 e^{\lambda t}\rho(\lambda)d\lambda}{\int e^{\lambda t}\rho(\lambda)d\lambda}. \qquad (6.22)$$

And if ϵ is fixed (i.e., ϵ does not depend on N),

$$T^{(1)}(H) \approx \inf\{t : E_{\infty}(t) \le \epsilon^2\}.$$

We arrive at an interesting dichotomy:

(1) If $\epsilon \le N^{-5/3 - \sigma/2}$ then Theorem 6.2 gives the approximation

$$T^{(1)}(H) = \frac{\log \epsilon^{-1} - \frac{2}{3}\log N}{\lambda_1 - \lambda_2}(1 + o(1)).$$

This is, to leading order, a random quantity that is a *local eigenvalue statistic*.

6.3 Adding Probability

(2) If ϵ is fixed then we have the approximation

$$T^{(1)}(H) = \inf\{t : E_\infty(t) \le \epsilon^2\} + o(1).$$

This is, to leading order, a deterministic quantity that is a *global eigenvalue statistic*.

An interesting problem is to understand how $T^{(1)}$ transitions from a local to global eigenvalue statistic as ϵ increases.

Recent results, see Ding and Trogdon (2021), allow t to grow logarithmically (with respect to N) in (6.21), which then allows ϵ to decrease similarly in case (2), with the conclusion remaining the same. Yet, there is still a gulf between (1) and (2).

References

Abraham, R, and Marsden, JE. 1978. *Foundations of Mechanics, Revised, Enlarged, Reset*. Benjamin/Cummings. 16, 24

Adler, M. 1979. Trace functional for formal pseudo-differential operators and the symplectic structure of the Korteweg-de Vries type equations. *Inventiones Mathematicae*, **50**, 219–248. 32

Agrotis, M, Damianou, P, and Sophocleous, C. 2006. The Toda lattice is super-integrable. *Physica A: Statistical Mechanics and Its Applications*, **365**, 235–243. 33, 112

Arnold, VI. 1978. *Mathematical Methods of Classical Mechanics*. Springer. 16

Bakhtin, Y, and Correll, J. 2012. A neural computation model for decision-making times. *Journal of Mathematical Psychology*, **56**(5), 333–340. 4

Beals, R, and Sattinger, DH. 1991. On the complete integrability of completely integrable systems. *Communications in Mathematical Physics*, **138**(3), 409–436. 30

Bloemendal, A, Erdős, L, Knowles, A, Yau, H-T, and Yin, J. 2014. Isotropic local laws for sample covariance and generalized Wigner matrices. *Electronic Journal of Probability*, **19**, 1–53. 150

Bourgade, P. 2016. *Personal communication*. 132

Bourgade, P, and Yau, H-T. 2017. The Eigenvector Moment Flow and local Quantum Unique Ergodicity. *Communications in Mathematical Physics*, **350**(1), 231–278. 122, 125, 126

Bourgade, P, Erdős, L, and Yau, H-T. 2014. Edge universality of beta ensembles. *Communications in Mathematical Physics*, **332**(1), 261–353. 123, 125, 132

Chu, MT. 1984. The generalized Toda flow, the QR algorithm, and the Center Manifold Theory. *SIAM Journal on Algebraic Discrete Methods* **5**(2), 187–201. 7

Chu, MT. 1986. A differential equation approach to the singular value decomposition of bidiagonal matrices. *Linear Algebra and Its Applications*, **80**, 71–80. 14

Date, E, and Tanaka, S. 1976. Analogue of inverse scattering theory for the discrete Hill's equation and exact solutions for the periodic Toda lattice. *Progress of Theoretical Physics*, **55**(2), 457–465. 13

Deift, P. 2000. *Orthogonal Polynomials and Random Matrices: A Riemann-Hilbert Approach*. American Mathematical Society. 122

Deift, P, and Gioev, D. 2007. Universality at the edge of the spectrum for unitary, orthogonal, and symplectic ensembles of random matrices. *Communications on Pure and Applied Mathematics*, **60**(6), 867–910. 125

Deift, P, and Trogdon, T. 2017. Universality for Eigenvalue algorithms on sample covariance matrices. *SIAM Journal on Numerical Analysis*, **55**(6), 2835–2862. 5, 11, 133

Deift, P, and Trogdon, T. 2018. Universality for the Toda algorithm to compute the largest Eigenvalue of a random matrix. *Communications on Pure and Applied Mathematics*, **71**(3), 505–536. 5, 11

Deift, P, Kamvissis, S, Kriecherbauer, T, and Zhou, X. 1975. The Toda rarefaction problem. *Communications on Pure and Applied Mathematics*, **49**(1), 35–83. 13

Deift, P, Lund, F, and Trubowitz, E. 1980. Nonlinear wave equations and constrained harmonic motion. *Communications in Mathematical Physics*, **74**(2), 141–188. 24, 26, 102

Deift, P, Nanda, T, and Tomei, C. 1983. Ordinary differential equations and the symmetric Eigenvalue problem. *SIAM Journal on Numerical Analysis*, **20**, 1–22. 3, 7, 10

Deift, P, Li, L-C, Nanda, T, and Tomei, C. 1986. The Toda flow on a generic orbit is integrable. *Communications on Pure and Applied Mathematics*, **39**(2), 183–232. 33, 66, 71, 75, 88, 96, 100

Deift, P, Kriecherbauer, T, McLaughlin, KT-R, Venakides, S, and Zhou, X. 1999. Uniform asymptotics for polynomials orthogonal with respect to varying exponential weights and applications to universality questions in random matrix theory. *Communications on Pure and Applied Mathematics*, **52**(11), 1335–1425. 124

Deift, PA, Li, L-C, Spohn, H, Tomei, C, and Trogdon, T. 2022. On the open Toda chain with external forcing. *Pure and Applied Functional Analysis*, **7**(3), 915–945. 8, 14

Deift, PA, Menon, G, Olver, S, and Trogdon, T. 2014. Universality in numerical computations with random data. *Proceedings of the National Academy of Sciences of the United States of America*, **111**(42), 14973–14978. 3, 4, 132

Ding, X, and Trogdon, T. 2021. A Riemann–Hilbert approach to the perturbation theory for orthogonal polynomials: Applications to numerical linear algebra and random matrix theory. *arXiv preprint 2112.12354*, 1–77. 151

Dunford, N, and Schwartz, JT. 1988. *Linear Operators, Part I: General Theory*. Wiley. 69

Egorova, I, Michor, J, Pryimak, A, and Teschl, G. 2023. Long-time asymptotics for Toda shock waves in the modulation region. *Journal of Mathematical Physics, Analysis, Geometry*, **19**(1), 396–442. 13

Erdős, L. 2012. Universality for random matrices and log-gases. *Current Developments in Mathematics*, **2012**(1), 59–132. 126

Erdős, L, Yau, H-T, and Yin, J. 2012. Rigidity of Eigenvalues of generalized Wigner matrices. *Advances in Mathematics*, **229**(3), 1435–1515. 122, 126

Erdős, L, Knowles, A, Yau, H-T, and Yin, J. 2013. The local semicircle law for a general class of random matrices. *Electronic Journal of Probability*, **18**, 1–58. 122

Flaschka, H. 1974a. On the Toda lattice. II: Inverse-scattering solution. *Progress of Theoretical Physics*, **51**(3), 703–716. 8, 13, 32, 37

Flaschka, H. 1974b. The Toda lattice. I. Existence of integrals. *Physical Review B*, **9**(4), 1924–1925. 8, 13, 32, 37

Francis, J. 1961. The QR transformation: A unitary analogue to the LR transformation - Part 1. *The Computer Journal*, **3**(4), 265–271. 9

Fried, D. 1986. The cohomology of an isospectral flow. *Proceedings of the American Mathematical Society*, **98**, 363–368. 121

Gaifullin, A. 2009. The manifold of isospectral symmetric tridiagonal matrices and realization of cycles by aspherical manifolds. *Proceedings of the Steklov Institute of Mathematics*, **263**, 38–56. 121

Gardner, CS, Greene, JM, Kruskal, MD, and Miura, RM. 1967. Method for solving the Korteweg–de Vries equation. *Physical Review Letters*, **19**, 1095–1097. 31, 32

Golub, GH, and Van Loan, CF. 2013. *Matrix Computations*. Johns Hopkins University Press. 3

Gragg, WB, and Harrod, WJ. 1984. The numerically stable reconstruction of Jacobi matrices from spectral data. *Numerische Mathematik*, **44**(3), 317–335. 43

Jiang, T. 2006. How many entries of a typical orthogonal matrix can be approximated by independent normals? *The Annals of Probability*, **34**(4), 1497–1529. 125

Kac, M, and Van Moerbeke, P. 1975. A complete solution of the periodic Toda problem. *Proceedings of the National Academy of Sciences of the United States of America*, **72**(8), 457–465. 13

Kappeler, T, and Pöschel, J. 2003. *KdV&KAM*. Springer. 33

Kato, T. 1995. *Perturbation Theory for Linear Operators*. Springer. 41

Kirillov, A. 2004. *Lectures on the Orbit Method*. AMS. 16, 24, 67

Knowles, A, and Yin, J. 2017. Anisotropic local laws for random matrices. *Probability Theory and Related Fields*, **169**(1–2), 257–352. 150

Kodama, Y, and Shipman, B. 2018. Fifty years of the finite nonperiodic Toda lattice: a geometric and topological viewpoint. *Journal of Physics A: Mathematical and Theoretical*, **51**353001. 8

Kostant, B. 1979. The Solution to a generalized Toda Lattice and representation theory. *Advances in Mathematics*, **34**(4), 195–338. 7

Lax, PD. 1975. Periodic solutions of the KdV equation. *Communications on Pure and Applied Mathematics*, **28**, 141–188. 32

Leite, R, Saldanha, S, and Tomei, C. 2008. An atlas for tridiagonal isospectral manifolds. *Linear Algebra and Applications*, **429**, 387–402. 51, 104, 113, 117, 118

Leite, R, Saldanha, S, and Tomei, C. 2010. The asymptotics of Wilkinson's shift: Loss of cubic convergence. *Foundations of Computational Mathematics*, **10**, 15–36. 34, 118

Leite, R, Saldanha, S, and Tomei, C. 2013. Dynamics of the symmetric Eigenvalue problem with shift strategies. *International Mathematics Research Notices*, **2013**, 4382–4412. 34, 118

Leite, R, Saldanha, S, Tomei, C, and Torres, DM. 2023. Linearizing Toda and SVD flows on large phase spaces of matrices with real spectrum. *Physics B*, **450**. Paper no. 133752, 10 pp. 14, 104, 113, 118

Li, L-C. 1986. SVD flows on generic symplectic leaves are completely integrable. *Advances in Mathematics*, **128**, 82–118. 14

Manakov, SV. 1974. Complete integrability and stochasticization of discrete dynamical systems. *Soviet Physics JETP*, **40**(2), 269–274. 8, 13, 32, 37

Manakov, SV. 1976. Remarks on the integration of the Euler equations of an n-dimensional rigid body. *Functional Analysis and Applications*, **10**(4), 328–329. 33

Monthus, C, and Garel, T. 2013. Typical versus averaged overlap distribution in spin glasses: Evidence for droplet scaling theory. *Physical Review B*, **88**(13), 134204. 126

Moser, J. 1975a. Finitely many mass points on the line under the influence of an exponential potential – an integrable system. *Dynamical Systems, Theory and Applications*. Lecture Notes in Physics, vol. 38. Springer, pp. 467–497. 9

Moser, J. 1975b. Three integrable Hamiltonian systems connected with isospectral deformations. *Advances in Mathematics*, **16**(2), 197–220. 7, 46

Moser, J. 1981. *Integrable Hamiltonian Systems and Spectral Theory*. Lezioni Fermiane. Accademia Nazionale dei Lincei. 24, 26

Moser, J, and Zehnder, EJ. 2005. *Notes on Dynamical Systems*. AMS. 16, 28, 62

Perret, A, and Schehr, G. 2014. Near-extreme Eigenvalues and the first gap of Hermitian random matrices. *Journal of Statistical Physics*, **156**(5), 843–876. 126, 132

Pfrang, CW, Deift, and Menon, G. 2014. How long does it take to compute the Eigenvalues of a random symmetric matrix? *Random Matrix Theory, Interacting Particle Systems, and Integrable Systems, MSRI Publications*, **65**, 411–442. 1, 3

Ramírez, JA, Rider, B, and Virág, B. 2011. Beta ensembles, stochastic Airy spectrum, and a diffusion. *Journal of the American Mathematical Society*, **24**(4), 919–944. 125, 127

Reed, M, and Simon, B. 1978. *Analysis of Operators*, vol. IV. Academic Press. 41

Reyman, AG, and Semenov-Tian-Shansky, MA. 1979. Reduction of Hamiltonian systems, affine Lie algebras and Lax equations. *Inventiones Mathematicae*, **54**(1), 81–100. 75

Shcherbina, M. 2009. Edge universality for orthogonal ensembles of random matrices. *Journal of Statistical Physics*, **136**(1), 35–50. 125

Shub, M, and Vasquez, A. 1987. Some linearly induced Morse-Smale systems, the QR algorithm and the Toda lattice. *Contemporary Mathematics*, **64**, 181–194. 118

Soshnikov, A. 1999. Universality at the edge of the spectrum in Wigner random matrices. *Communications in Mathematical Physics*, **207**(3), 697–733. 125

Stam, AJ. 1982. Limit theorems for uniform distributions on spheres in high-dimensional Euclidean spaces. *Journal of Applied Probability*, **19**(1), 221–228. 125

Symes, WW. 1980. Hamiltonian group actions and integrable systems. *Physica 1D*, **4**(1), 339–374. 10, 75

Symes, WW. 1982. The QR algorithm and scattering for the finite nonperiodic Toda lattice. *Physica D: Nonlinear Phenomena*, **4**(2), 275–280. 10, 66, 75, 77

Tao, T, and Vu, V. 2010. Random matrices: Universality of local Eigenvalue statistics up to the edge. *Communications in Mathematical Physics*, **298**(2), 549–572. 125

Tomei, C. 1984. The topology of isospectral manifolds of tridiagonal matrices. *Duke Mathematical Journal*, **51**, 981–996. 112, 121

Torres, D, and Tomei, C. 2013. An Atlas adapted to the Toda flow. *International Mathematics Research Notices*, **2013**(16), 13867–13908. 104, 113, 118

Tracy, CA, and Widom, H. 1994. Level-spacing distributions and the Airy kernel. *Communications in Mathematical Physics*, **159**, 151–174. 125

Venakides, S, Deift, P, and Oba, R. 1991. The Toda shock problem. *Communications on Pure and Applied Mathematics*, **44**(8–9), 1171–1242. 13

Warner, FW. 1983. *Foundations of Differentiable Manifolds and Lie Groups*. Graduate Texts in Mathematics, vol. 94. Springer. 16, 21

Watkins, DS. (1984). Isospectral flows. *SIAM Review*, **26**(3), 379–391. 7

Witte, NS, Bornemann, F, and Forrester, PJ. 2013. Joint distribution of the first and second Eigenvalues at the soft edge of unitary ensembles. *Nonlinearity*, **26**(6), 1799–1822. 126, 133, 134

Notations and Abbreviations

$E_\infty(t)$		150
β_j		123
$\mathcal{O}_{\Lambda,p}$	isospectral manifold	111
Σ^π		106
$\mathcal{S}_p, \mathcal{Z}_p^\pi, \mathcal{S}_p^\pi$		111
δ_j		127
D^π		105
$\ell(T^*)$		143
$\varepsilon(\sigma)$	parity of the permutation σ	54
γ_k, ν_k		94
Γ_N	$N \times N$ Hermitian matrices	73
γ_n		123
$\hat{\mathfrak{l}}_c^c$		143
$\hat{T}$		146
$\hat{T}_\gamma$		135
$\hat{\theta}_k$		95
$\hat{v}_{r,k}$	left generalized eigenvectors	99
Λ_2	second exterior power of $\mathbb{C}^N$	82
$\langle \cdot, \cdot \rangle$	standard inner product	17
$\mathfrak{l}_0, \mathfrak{l}_0^0$	lower triangular algebras	66
$\mathcal{J} = \mathcal{J}_N$	$N \times N$ Jacobi matrices	42
μ_k	angles for Toda on Jacobi matrices	88
μ_k, ν_k		86
μ_N		123
$\mu_{N,t}(d\lambda)$		150
ν_n		128
$\pi_{k^\perp}, \pi_{\mathfrak{l}_0^\perp}$	projections	66
$\pi_{s\ell}, \tilde{\pi}_u$	complementary projections	109

$\Re z$	real part of $z \in \mathbb{C}$	73
$\rho(x)$		123
$\rho_{N,k}(S)$		138
Σ_N	$N \times N$ symmetric matrices	1
$\mathrm{spec}(M), \sigma(M)$	the spectrum of M	1
$\tilde{\lambda}_i$		91
Up, Up^+	upper triangular groups	105
V_p	subspace of matrices with profile p	110
$\widetilde{B}(X)$	skew-Hermitian part of X	73
$\{\lambda_{jk}\}$	generalized eigenvalues	96
a_N		146
$A_N, U_N, D_N, \tilde{U}_N, \tilde{D}_N$	sets of special matrices	80
a_V, b_V, c_V		123
$B(X)$	skew-symmetric part of X	63
C_μ		123
D_k		52
$E(t)$		11
$E_0(t), E_1(t)$		137
$E_{1k}/E_{0,k}$	coadjoint invariants	97
$E_{r,k}$		96
F^{gap}		11
$F_\beta^{\mathrm{gap}}(t)$		126
$f_{rk}(X), \mu_{rk}(X)$		99
$G_{N,p}$		125
H_{Toda}	Toda Hamiltonian	8
I_c		127
L_α		140
Lo, Lo^1	lower triangular groups	66
$R_{N,s}$		125
S_N	permutation group on N symbols	106
$SO, SO(N)$	special orthogonal group	106
T^*		142
$T^{(1)}$		129
T_2	extension of T to Λ_2	83
T_{ODE}		133
$v_{12,jk}$	exterior product entry	82
$v_{r,k}$	right generalized eigenvectors	96
$w_{r1}, \hat{w}_{r1}$		99
X_k		96
x_{cm}, y_T	center of mass and momentum of Jacobi Toda	90

GUE, BUE, GOE, BOE 133
$M_N(\mathbb{R})$ 22
WE, IE 122

Index

action-angle variables, 29
  for Jacobi matrices, 86
  on full symmetric matrices, 96
Adler–Kostant–Symes theorem, 32

band-preserving flow, 39

co adjoint invariants, 97
co adjoint orbits, 20
conjugate gradient, 3
constrained systems, 20
cotangent bundles, 20

deflation, 2
deflation time, 2
deflation time, normalized, 3

eigenvalue statistic, global, 151
eigenvalue statistic, local, 150
Euler–Arnold equation, 32
exponentially small error, 56

flow, 1

generalized eigenvalue problem, 98
generalized eigenvalues and eigenvectors, 96
generic orbits, 96
generic symmetric matrix, 96
genetic algorithm, 4
GMRES, 3

halting time, 11
Hamiltonian mechanics, 16
Hamiltonians, 16
harmonic oscillator, 34
history of the Toda system, 8

integrable systems, 16
inverse scattering methods, 13
isospectral manifold, 111
isospectrality, 1

Jacobi eigenvalue algorithm, 3

Jacobi identity, 18
Jacobi identity for Lie algebras, 21
Jacobi matrix, 14
Jacobi matrix, complex, 74

Killip's construction, 31
Korteweg–de Vries equation, 32
Korteweg–de Vries equation, periodic, 33
Kostant–Kirillov form, 20

Lax pair, 14
Lax–Levermore–Venakides method, 13
lexicographically ordered basis, 82
Lie–Poisson structure, 15
Liouville–Arnold–Jost (LAJ) integrability, 27
local laws, 150

matrix
  π-admissible, 106
  LU-positive decomposition, 105
  sign diagonal, 106
  upper Hessenberg, 110
measure, averaged empirical spectral, 123
measure, equilibrium, 123

Newton's cradle, 50

ordering eigenvalue algorithm, 48

perturbation theory, 41
phase shifts, 51
Poisson bracket, 17
Poisson manifolds, 19
power methods, direct and inverse, 133
principal minors, 105
profile, 110

QR, $\hat{Q}L$ factorizations, 15
QR algorithm, 9
quadrant, 110
quadrature, 27

Index

quantiles, 124
random matrix ensembles, invariant, 11, 122
random matrix ensembles, Wigner, 11, 122
Riemann–Hilbert/steepest descent method, 13
rigid body in N dimensions, 33
rigidity, 124
scaling regime, 12
scaling region, 11, 130
simple eigenvalue, algebraic and geometric, 41
simple pendulum, 34
skew-symmetric tensors, 51
spectral measure, 52
spectrum, 1
stochastic Airy operator, 125
stroboscope property, 10
super-integrability, 33, 112
SVD flows, 14
Symes QR factorization method, 33
symmetric matrix, 1
symplectic leaves, 24
symplectic manifolds, 16
symplectic map, 26
symplectomorphism, 28
Toda algorithm, 7
Toda equation, generalized, 62
Toda equation, Hermitian case, 73
Toda flow, 5
Toda flow, generalized extended, 75
Toda lattice
  with external forcing, 14
  infinite number of particles, 13
  open, 13
  periodic, 13
Toda rarefaction problem, 13
Toda shock problem, 13
Toda system, open, 8
tridiagonal matrix, 38
true error, 148
Z-variables, 14

For EU product safety concerns, contact us at Calle de José Abascal, 56–1°, 28003 Madrid, Spain or eugpsr@cambridge.org.

www.ingramcontent.com/pod-product-compliance
Ingram Content Group UK Ltd.
Pitfield, Milton Keynes, MK11 3LW, UK
UKHW022332300126
467541UK00021B/409